Assessing Calculus Reform Efforts

A Report to the Community

Assessing Calculus Reform Efforts
Advisory Committee

Professor Ronald G. Douglas
Office of the Vice Provost
SUNY at Stony Brook
Stony Brook, NY 11794

Professor Wade Ellis, Jr.
Department of Mathematics
West Valley College
14000 Fruitvale Avenue
Saratoga, CA 95070

Professor Judy Franz
Department of Physics
University of Alabama in Huntsville
Huntsville, AL 35899

Professor James T. Fey
Department of Mathematics
University of Maryland
College Park, MD 20742

Professor Pamela Ferguson
Office of the President
Grinnell College
Grinnell, IA 50112

Professor James R. C. Leitzel
Project Director
Department of Mathematics and Statistics
University of Nebraska-Lincoln
Lincoln, NE 68588

Professor Robert E. O'Malley, Jr.
Dept. of Applied Mathematics, FS-20
University of Washington
Seattle, WA 98195

Professor Alan C. Tucker
Chair (1992–1994)
Department of Applied Mathematics
SUNY at Stony Brook
Stony Brook, NY 11794

Professor Lynn A. Steen
Chair (1991)
Department of Mathematics
St. Olaf College
Northfield, MN 55057

Evaluation Consultant
Professor John Dossey
Department of Mathematics
Illinois State University
Normal, IL 61761

Administrative Assistant
Ms. Jane Heckler
The Mathematical Association of America
1529 Eighteenth Street, NW
Washington, DC 20036

Assessing Calculus Reform Efforts

A Report to the Community

Alan C. Tucker
SUNY at Stony Brook

James R. C. Leitzel
University of Nebraska-Lincoln

Published by
The Mathematical Association of America

This project was supported, in part, by the National Science Foundation.
Opinions expressed in this report are those of the authors and not necessarily
those of the foundation.

MAA Notes and Reports Series

The MAA Notes and Reports Series, started in 1982, addresses a broad range of topics and themes of interest to all who are involved with undergraduate mathematics. The volumes in this series are readable, informative, and useful, and help the mathematical community keep up with developments of importance to mathematics.

MAA Notes

MAA Reports

These volumes may be ordered from:
The MAA Service Center
P.O. Box 90973
Washington, DC 20036
800-331-1MAA   FAX 301-206-9789

Acknowledgements

The preparation of this report on the state of involvement of mathematical sciences departments (their faculty and students) in efforts to revise calculus instruction required the willing cooperation of many individuals. We would like to express our sincere appreciation to those in institutions across the country who took the time to complete the various survey forms, engaged in conversations with us about their efforts, and willingly shared their own enthusiasm about the enhanced learning environments change has fostered. Special thanks are accorded the directors of various projects in calculus reform. They provided valuable insightful information about the philosophy and goals of their own projects and the movement as a whole.

We particularly recognize the contributions to this report made by John A. Dossey, Illinois State University. John provided critical review of the forms and compiled and analyzed the data from the various surveys the project conducted. The advice and suggestions of the Advisory Committee were instrumental in creating the framework for the assessment. The Division of Undergraduate Education at the National Science Foundation, especially program officers James Lightbourne and William Haver, provided necessary detailed information that enabled us to have a broader perspective about the Calculus Initiative as it was implemented at the Foundation.

However, the real significance of the calculus reform effort is the change that is occurring in collegiate classrooms across the country. Faculty and students are enthusiastic about the learning environments that the new approaches are creating. The future is bright.

Alan C. Tucker, SUNY at Stony Brook
James R. C. Leitzel, University of Nebraska–Lincoln

Contents

Executive Summary

Calculus reform is a growing endeavor. A survey of mathematics departments conducted in the spring of 1994 found that 22% of 1048 responding mathematics departments were engaged in major reform efforts, with another 46% reporting modest efforts under way. There is every indication that calculus reform is gaining widespread, though far from universal, acceptance.

This report presents the findings of a Mathematical Association of America (MAA) study assessing the calculus reform movement. The assessment involves both quantitative measurements of change, obtained from a set of short and in-depth surveys, and qualitative judgments based on presentations and discussions at numerous calculus conferences, workshops, and contributed-paper sessions at MAA national and sectional meetings, as well as conversations with scores of collegiate mathematics leaders and individual mathematics instructors.

The number of students currently enrolled in reform calculus sections at the 600 institutions responding to the request for student data in the spring 1994 survey was approximately 125,000. This provides a very conservative lower bound on reform enrollments in spring 1994. However, other information suggests that, extrapolating from this number, a more reasonable estimate would be at least 150,000 students in spring 1994, about 32% of all students then taking calculus. Many of the institutions that currently report using reform materials in experimental sections have plans to move their efforts to course-wide adoption in the near future.

A key finding of the assessment study has been that *how* calculus is taught has changed more than *what* is taught. From the start of the reform movement, there has been broad agreement that a guiding theme should be to concentrate on greater conceptual understanding, developed through extensive numerical, graphical, algebraic and modeling interpretations. Changes include the use of graphing calculators and computers, open-ended projects, extensive writing, more applications, and use of cooperative learning groups. Institutions report ongoing, lively discussions among faculty about approaches to topical material, modes of instruction, and ways of assessing students' knowledge. Many faculty are concerned with getting the "right mix" of hands-on and technology-assisted work. The lecture method is being questioned and modified and, in some classrooms, abandoned.

Large numbers of reform instructors report that the new instructional methods are having positive effects on students' conceptual understanding, mathematical reasoning, and problem-solving abilities. In a significant number of cases, instructors report noticeable improvements in retention and passing rates. The hallmarks of calculus reform—changes in modes of instruction and use of technology along with an increased focus on conceptual understanding and decreased attention to symbol manipulation—are finding their way into both pre-calculus and post-calculus mathematics courses. Enrollments in post-calculus courses seem to be improving. Students' association of heavy technology usage in some calculus experiences has created a positive image for mathematics.

Faculty concerns have centered around the extra time needed to prepare for non-traditional instruction and to develop and read student projects. All agree that teaching a "reformed" course the first time takes a great deal of time and effort—and humility. After teaching reform calculus a few times, the preparation time is reduced but is still more than in traditional courses. Information about the response of client disciplines to calculus reform is spotty since reform is just beginning at most institutions, although feedback has been generally positive. In particular, facility with numerical approaches has frequently been cited as a major benefit of calculus reform.

In spring 1993 and spring 1994, the study obtained responses to an in-depth survey from 62 mathematics departments involved in some level of calculus reform. This assessment study also undertook two short surveys in spring 1992 and spring 1994 which generated 745 and 1067 responses, respectively. In the 1992 short survey, 56% of responding departments indicated modest or major change was occurring. In the 1994 short survey, 68% of responding departments indicated that same level of change. More detailed information by level of institution

1994 Short Survey: Response by Institution Degree Type and Level of Change
(Percentages in parentheses are with respect to numbers in each row)

Highest degree /Change	None	Modest	Major	Total Responses	Total Institutions
Doctoral	36 (24%)	82 (55%)	32(21%)	150	165
Master's	35 (26%)	67 (50%)	33 (24%)	135	236
Bachelor's	116 (26%)	207 (47%)	120 (27%)	443	1020
Associate	147 (46%)	126 (39%)	47 (15%)	320	1018
BS + MS + PhD	187 (26%)	356 (49%)	185 (25%)	728	1421
Total	334 (32%)	482 (46%)	232 (22%)	1048	2439

is contained above (Table 3 from the report).

The in-depth survey indicated that about half of current reform efforts involve a traditional text supplemented by reform materials. The remaining efforts involve the use of a reform text. The Harvard Consortium text is used by the majority of institutions that have selected a reform text (315 in fall 1993 and over 500 in fall 1994). It is worth noting again that most institutions using reform materials in some experimental sections report they expect to move to course-wide adoption in the near future.

The report examines some of the historical developments that set the stage for the current reform movement. It notes that most of the elements that characterize the calculus reform movement were proposed by an MAA committee in 1953. More recently, the 1986 Tulane Calculus Workshop set the stage for the current calculus reform efforts by documenting the shortcomings of existing calculus instruction, by outlining themes in the design of reformed calculus courses, and by proposing a model strategy for creating, disseminating, and implementing reform. The Tulane Workshop report provided a blueprint for reform that has been widely accepted. That blueprint has given a solid foundation and invaluable coherency to the reform movement.

Faculty responding to the project's surveys indicated that reform teaching takes more time in grading papers, meeting with students, dealing with technology, and planning for class. Developing a reform course takes a tremendous amount of time. But there was a strong feeling of growth and professionalism among faculty, particularly at non-doctoral institutions. There has been increased discussion of teaching and learning among the

faculty. They are experimenting with other modes of evaluation of student learning besides standard paper-pencil timed tests.

From the start of the reform movement, the National Science Foundation (NSF) and the mathematical community put a heavy emphasis on dissemination. Project investigators, committees of the Mathematical Association of America, and other independent groups organized efforts to publicize various funded and non-funded approaches to calculus reform. There is a spectrum of levels of reform taking place, and faculty attitudes towards reform are just as diverse. While some faculty might resist reform at the level of the Harvard Consortium text, other faculty argue that the Consortium text is so modest in its goals that it does not constitute reform at all. Not surprisingly, the vast majority of reform courses being offered are at the less radical level.

The reform movement in calculus appears to be spreading to other aspects of undergraduate mathematics programs. Many respondents to the project's surveys indicated that they were looking to implement a graphical/functional viewpoint in precalculus and that technology was being introduced into other courses, most commonly precalculus, linear algebra, and differential equations. At the bachelor's and master's institutions, there were reports of the use of computer algebra systems (CAS) in upper division courses as a result of their usefulness in the calculus. After using CAS in calculus, one respondent said, "I cannot wait to teach complex analysis next year using CAS." There were also reports of greater use of discussion versus lecture in the classes. A few doctoral-level institutions reported that change occurred first in differential equations and linear algebra and then moved down to calculus.

Calculus reform has also encouraged increased interest in research about how undergraduate students learn mathematics. At present, there is only modest evaluation data on the effectiveness of the various reform courses and teaching strategies. The promising start of the calculus reform movement has led to calls for increased research and educational yardsticks to evaluate learning in calculus, in both traditional and reform courses, in terms of the larger goals of a mathematical education. A special NSF workshop was convened in July 1992 whose purpose might informally be summarized as, "If calculus reform is succeeding, what is it that students should be learning and how will we know?" The report from that workshop is available from the NSF.

There is no question of the importance the NSF initiative has had in achieving the changes reported to date. The NSF program successfully directed the mathematics community to address the task of reforming the calculus curriculum and provided coherence to those efforts. Now, the community is beginning to reexamine other parts of the undergraduate curriculum as well. The new initiative proposed by the NSF Division of Undergraduate Mathematics Education, *Mathematical Sciences and their Applications Throughout the Curriculum,* extends the efforts of the calculus reform movement by encouraging broad interaction among many disciplines.

Some comments by individuals speak more forcefully to the role of NSF:

> There is absolutely no doubt that the NSF Calculus Initiative was fundamental to the changes that have taken place in our program. Quite frankly we had no idea how complex the task of changing our culture would be and no idea of how much time and energy it would consume. The NSF program supported a number of models and materials development programs. From some of these (e.g. St. Olaf) we have drawn extensively, from others (unnamed) we have seen what to avoid. However all of them have been useful. Finally, the financial support the program provided has been pivotal to actually getting the changes made. Without it we simply would not have had the resources to carefully adapt the extant models to one that best suits our situation.

> Absolutely, because without it I doubt whether the movement would have been able to build the momentum it has now produced. I am a bit worried about the use of the past tense in reference to the NSF calculus initiative, however. The job of reform is FAR from finished, and it is to be hoped that the program continues!

> Yes, major funding has resulted in major change here, reaching far beyond calculus. Now the changes in calculus are affecting much of our undergraduate and graduate teaching. I dare say there has been more change here than at any other research department in the country, in terms of overall impact on the largest portion of faculty.

This report presents results of calculus reform efforts through spring 1994. The surveys conducted by the project did not ask respondents to predict whether the movement has been a success or whether it would be universally implemented in some form in all institutions by the year 2000. However, there is a sentiment growing in many departments not now fully converted to reform, that reform is becoming inevitable. The fact that in the past two years over 95% of institutions using a reform text continue using a reform text the next year (for at least some sections) indicates that calculus reform is likely to be around for the foreseeable future.

Support for this assessment study was provided, in part, by the National Science Foundation. However, this report was developed independently of the NSF, and opinions expressed in the report are those of the authors and not necessarily those of the Foundation.

Introduction

Calculus reform is taking place at all levels of post-secondary institutions. In a spring 1994 survey, 68% of 1048 responding mathematics departments indicated that either modest or major calculus reform efforts were currently under way. Of the departments responding, 22% were undertaking major reform efforts and 46% described their efforts as modest. We estimate that at least 150,000 students, or 32% of all calculus enrollments, in spring 1994 were in reform courses.

In many institutions, the change at present is modest—an experimental section using a reform text or a course-wide de-emphasis of integration techniques accompanied by the introduction of graphing calculators. In others, the change is quite pronounced, with course-wide use of a reform text, graphing calculators or computers, cooperative learning, open-ended projects, regular writing assignments, or increased emphasis on modeling and applications. Faculty report working much harder to prepare for such reformed classes but overwhelmingly indicate their strong feeling that the extra effort is worth it, even though some have seen few signs to date of meaningful changes among their students. At the same time, at several institutions there are significant numbers of faculty who remain skeptical of the claims of calculus reformers and who fear reform efforts will only make a bad situation worse.

Institutions report ongoing, lively discussions among faculty about approaches to the topical material, modes of instruction, and ways of assessing students' knowledge. Many faculty are concerned about getting the "right mix" of hands-on and technology-assisted work. The lecture method is being questioned and modified and, in some classrooms, abandoned. Large numbers of reform instructors report that new instructional methods are having positive effects on students' conceptual understanding, mathematical reasoning, and problem-solving abilities. In a significant number of cases, instructors report noticeable improvements in retention and passing rates. The hallmarks of calculus reform—changes in modes of instruction and use of technology, along with an increased focus on conceptual understanding and decreased attention on symbol manipulation—are finding their way into both pre- and post-calculus courses. The use of technology, while initially meeting resistance, is proving quite popular with students.

Information about the response of client disciplines to calculus reform is spotty since reform is just beginning at most institutions, although feedback has been generally positive. In particular, facility with numerical approaches has frequently been cited as a major benefit of calculus reform.

Faculty concerns have been centered around the extra time needed to prepare for non-traditional instruction and to develop and read student projects. All agree that teaching a "reformed" course the first time takes a great deal of time and effort—and humility. After teaching reformed calculus a few times, preparation time is reduced but is still much more than in traditional courses. The number of faculty who abandon reform approaches after trying them is very small. Likewise, over 95% of institutions that have tried reform texts one year continue using reform texts the following year. This persistence of reform among faculty and text orders indicates that calculus reform seems to be here to stay.

The causes of calculus reform are even more diverse than the forms it is taking. The National Science Foundation (NSF) funding has been the most visible and important source of support for calculus reform. The amount of money spent by NSF has been modest, averaging about $3,000,000 a year for seven years. In contrast, higher education expenditures for faculty salary and overhead for calculus instruction is estimated at over $500,000,000 a year.

However, the impact of this NSF support in seeding individual projects, in nurturing dissemination and implementation efforts, and in leveraging local support has been critical to the reform movement.

Another factor pushing the cause of calculus reform is the end of the Cold War and the changing national focus to the kind of economic competitiveness that requires a well-trained work force. This has led to a national spotlight on the poor educational achievement of United States students and an associated concern that research universities, the flagships of higher education, should give more attention to undergraduate teaching. The *Standards* produced by the National Council of Teachers of Mathematics (NCTM), the continuing concern at individual institutions about high failure rates in calculus, and the availability of moderately priced graphing calculators have also served to impel reform in calculus teaching. The reform movement has benefited from effective advocates. People such as Ron Douglas, Lynn Steen, and Deborah Hughes-Hallett have effectively articulated and championed the need to change how calculus is taught. However, the reform movement has been largely a grassroots effort with reform developed and implemented on a local level.

With support from the National Science Foundation, the Mathematical Association of America (MAA) has conducted an assessment of the nationwide calculus reform effort. The task was to review various aspects of the movement to get an assessment of the current attitudes and involvement of mathematical sciences departments—their faculty and students—in efforts to revise instruction in calculus. This study *Assessing Calculus Reform Efforts*, was not intended to assess outcomes of individual projects or to undertake any "official" review of NSF's calculus program initiatives, but rather to provide a report on the movement as a whole.

Background

The Modernization of the Collegiate Mathematics Curriculum in the 1950s and 1960s

Readers knowledgeable about the history of American higher education are aware that most contemporary educational crises and reform movements address problems that have arisen several times in the past century. Moreover, the calculus reform begun in the late 1980s had its roots in major changes in collegiate mathematics in the 1950s and 1960s. Calculus has been taught for centuries, but it is only recently that it assumed its current role as a standard freshman mathematics course. In 1950, it was the norm for college mathematics to start with a year-long course in college algebra and trigonometry. Calculus with analytic geometry was a sophomore level course. At that time, an undergraduate engineering degree required 5 years of study.

Universal Mathematics Project. In 1953 the Mathematical Association of America created the Committee on the Undergraduate Program (CUP), a precursor to the current Committee on the Undergraduate Program in Mathematics (CUPM), in response to "widespread interest in revision of the elementary courses in mathematics." CUP developed plans for a course called *Universal Mathematics,* a year-long introduction to college mathematics for college freshmen, serving both engineers and humanists. A supplementary "laboratory" was proposed for engineers to develop technical skills in trigonometry, systems of equations, and such. The course assumed two rigorous years of high school algebra. (A course in Euclidean geometry was desirable but not required.) Because *Universal Mathematics* had several goals in common with the current calculus reform movement and because eminent mathematicians such as J. Kemeny, E.J. McShane, Saunders Mac Lane, A.W. Tucker, and S.S. Wilks were involved with its design, it is useful to review briefly the content and objectives of this effort to modernize freshman mathematics.

The course was divided into two parts. Our discussion will focus on Part I, "Functions and Limits," which was an introduction to calculus. (Part II, Mathematics of Sets, was developed by J. Kemeny and colleagues into the successful text, *Introduction to Finite Mathematics.*) The chapters in the "Functions and Limits" course were: coordinate systems; lines, vectors, proportion, and measurement theory; functions including informal max/min problems and divided differences; limits; polynomial differentiation with extensive applications; integrals; and logarithmic and exponential functions.

The text had a dual— almost schizophrenic— style of presentation, with parallel sections of intuitive presentation and formal (theoretical) presentation. The two approaches appeared on opposing pages. The informal sections gave extensive graphical and applied interpretations of functions and other calculus concepts. They used discrete (tabular) functions. For example, one of the interpretations of derivative was as the marginal cost in economics. Quoting from the text's preface, "there is some attempt to teach synthesis and problem solving in the spirit of Pólya's work. Moreover, we have tried to hold to the fundamental ideas of calculus within the framework of finite processes (difference quotients and Riemann sums) as long as possible. This procedure is intended both to make the approach easier for the student and also to emphasize numerical methods which are becoming ever more important."

On the other hand, the theoretical sections had a level of rigor that in places exceeded even what one would expect for an honors calculus course today, with axiomatic definition s of the plane and limit processes leading to theorems, such as "The final residue $\cap F$ of any limit process $F: N \rightarrow I_R$ is a closed interval." It was understood that the formal sections would not all be covered in most courses.

Both the formal and the interpretative approaches were quite revolutionary for their time. CUP recommended that this course should be followed in the sophomore year by a year-long, rigorous course in calculus with analytic geometry. The "Functions and Limits" course never made it beyond a preliminary text. As a result of pressures from engineering and physics departments and improving high school preparation in the mid-1950s a course in techniques of calculus became the standard mathematics for freshmen in engineering and physical sciences.

It is worthwhile noting that today the mathematics community is still trying to grapple with the problem of choosing among these three basic approaches to calculus instruction—technical, theoretical, or interpretative. Changing societal and educational concerns and the heterogeneous nature of college students suggest that any particular approach or mixture of approaches will work well for only a limited group of students for a limited period of time. By the late 1950s the country was engaged in the technological Cold-War battle for supremacy in space, and increased production of engineers was a high priority in the national agenda for academia. Not surprisingly, the "techniques" approach to calculus was the clear-cut winner in this period. The client departments liked it, and the students seemed to find it appropriately challenging and useful.

Modernization of the Undergraduate Program in Mathematics. Also in the 1950s, higher education enrollments were increasing rapidly. This growth created considerable concern in the mathematics community about the challenges of increasing the production of mathematics PhDs. The undergraduate major in mathematics had lagged behind research developments in the discipline, making the transition from undergraduate to graduate study traumatic, with a large number of first-year graduate students dropping out. In response to this problem, around 1960 the MAA charged its Committee on the Undergraduate Program in Mathematics (CUPM) to plan a modernized curriculum for the undergraduate mathematics program including more theory and course work in algebra and contemporary analysis so as to better prepare mathematics majors for graduate study. Extensive curriculum studies about the mathematics major and all undergraduate mathematics were undertaken by the MAA with liberal NSF support. Many of the leading research mathematicians of the time were heavily involved. With NSF funding, summer institutes were established to upgrade the mathematical knowledge of collegiate mathematics faculty, a minority of whom had a PhD in 1960. As part of this effort, honors calculus courses for potential mathematics majors were introduced that reflected the formal approach in *Universal Mathematics*. With the techniques approach being used for engineers

and the formal approach for mathematics majors, the interpretative approach of *Universal Mathematics*, with its conceptual visualization, applied interpretations, and discrete/numerical components, was almost totally neglected. The modernization of the undergraduate mathematics program in the 1960s is of interest to the calculus reform movement for another reason besides its promotion of technical and formal approaches to calculus. It was an example of how a variety of confluent interests led in a short period of time to a major reshaping of mathematics instruction. A striking example of this change is that, in less than a decade, the theory of vector spaces moved from a graduate-level course to a sophomore-level course. Key factors driving the successful modernization with a theoretical emphasis were:

- An undergraduate curriculum that had lagged behind the new analytical and algebraic framework for mathematics developed in the first half of the 20th century.

- A consensus among virtually all university and collegiate mathematicians about the need to update the out-of-date curriculum, and participation by all sectors of the community in reworking the curriculum.

- A greatly enhanced use of mathematics in engineering and science arising from technological advances initially spurred by World War II and later by the advent of digital computers.

- A societal demand for a large scientific enterprise, including foundational academic research, in support of the Cold War.

- An explosion of mathematical research activity fueled by European émigrés, the growth of doctoral training, and federal funding.

- A large cohort of solidly trained, well motivated potential mathematics majors (5% of the entering freshmen in 1965 expressed an interest in becoming mathematics majors; for the past 15 years, this number has been under 1%).

Mathematics in the Modern Technological World. At the start of the 20th century, America was a "frontier" country that valued initiative and entrepreneurial skills far above formal education. Technical creativity was epitomized by Thomas Edison, whose formal education ended at sixth grade. World War II, however, produced major societal and intellectual changes. The value of academic science and management was proved in the war effort. With the GI Bill available to millions of veterans, a college degree became both desirable and accessible to a large number of young Americans. The majority of college and university mathematics faculty did not have a PhD in 1960, although a small but growing number of faculty, seeded by talented European refugees, were rapidly advancing exciting research frontiers. Paralleling the growing value of a college education, a consensus quickly developed in academia that the PhD should be required for college teaching.

The modernization of the mathematics curriculum and professoriate coincided with the increased accessibility and importance of a college education and the growing role of mathematics in the modern world. The current calculus reform movement is seeking to address consequences, not fully anticipated at the time, of greater use of mathematics and a more diverse population of students that started in the 1950s and 1960s.

The "can do" spirit that mobilized the country for the Cold War effort was reflected in the streamlined mathematics sequence at the end of high school and beginning of college that allowed engineers to begin calculus and physics in the freshman year and obtain an engineering degree in just four years. The same sort of effort was sought in the mathematics community in the 1960s to enable mathematics undergraduates to be ready for graduate study in mathematics. The impetus to make the mathematics major more contemporary and theoretical was reinforced by parallel change occurring in mathematics faculty.

Although there may have been a swing a bit too far towards theory and preparation for graduate study, the reform of the mathematics major in the 1960s is impressive in the dramatic change in the undergraduate mathematics curriculum that resulted. It is worth noting that the theoretical thrust of the mathematics major was given an extra push in the late 1960s by politics. Specifically, the anti-war sentiment on campuses viewed the practical and applied aspects of mathematics as associated with the military/industrial complex. Pure mathematics was a safe retreat from this distasteful world.

The New Math. A discussion of mathematics education in the 1950s and 1960s is not complete without mention-

ing school mathematics. The educational changes discussed above also swept into school mathematics in the 1960s with the "New Math" movement. While initially motivated by a desire to update school curriculum in a balanced way, emphasizing set theory as much for its use in probability as for its role in theory, the "New Math" came to be oriented overwhelmingly towards theoretical foundations, with an absence of applications. The theoretical focus seen at the collegiate level was heightened in the schools by the large cohort of R. L. Moore students involved in the School Mathematics Study Group, the largest of the "New Math" efforts. Moore's intensive four-year mathematics honors program at the University of Texas was spectacularly successful in creating research mathematicians. However, the theoretical "New Math" materials produced a massive backlash when taught to large numbers of children by teachers frequently inadequately prepared for using these materials.

The backlash caused research mathematicians to retreat from involvement in school mathematics and, by extension, introductory college mathematics. The debacle of the "New Math" haunted school mathematics for decades and reactions to its failure at the collegiate level, as well as the school level, ensured that all mathematics reform efforts for many years, including calculus reform, would be highly sensitive to the needs of average students and involve extensive consultation with client groups. The public failure of the "New Math" stands in pointed contrast to the success of the reformed mathematics major which produced increased numbers of students ready and able to pursue doctoral training in mathematics. The reformed major program in collegiate mathematics focused itself toward building successful connections with graduate study rather than toward school mathematics.

The "New Math" did, however, exert influence on the K–12 curriculum during the 1970s and 1980s. At the elementary and middle school levels, there was a considerable expansion of the scope of what was known as mathematics. In fact, programs became mathematics programs rather than being centered in arithmetic. There were changes in secondary level curricular materials. For example, geometry courses shifted emphasis from the standard "two-column" proof, and there was more infusion of solid geometry into the curriculum. Thus in school mathematics the stage was set for the changes that were to come with the development of the NCTM *Standards.* In contrast to the 1960s, the development of the standards for school mathematics adopted the view that content, pedagogy, and context, as well as public relations, were all intertwined.

Setting the Stage for Calculus Reform in the 1970s and early 1980s

The 1960s Compact for Calculus Among Faculty, Students, and Client Disciplines Unravels. While mathematics was always essential for most scientific disciplines, in 1960 calculus was still viewed by many outside the physical sciences and engineering as a liberal arts subject. Some leading institutions in the 1960s required all students to take either three semesters of calculus or three semesters of a foreign language, somewhat in the same spirit that many college-bound students took Latin in high school. Through the 1970s and 1980s, calculus came to be recognized as essential knowledge for describing and explaining fundamental phenomena in many disciplines outside of the physical sciences. During this same period, as colleges enrolled a more diverse, less selective spectrum of students, college freshmen were enrolling with less mathematical preparation than in earlier decades. At the same time, many of the students who were better prepared and motivated took calculus in high school. The Advanced Placement Calculus Program of the College Board (AP) removed from college calculus courses a number of students who could be model learners and helpers to other students.

Upon seeing the apparent ease with which calculus reform has evolved recently, many people have asked why a reform effort was not started in the 1970s when problems first started to emerge. There are several reasons, foremost of which was the lack of adequate attention to calculus instruction in the mathematics profession. At major research universities, where mathematics faculty were prepared, the senior faculty had limited involvement with calculus instruction. They were preoccupied with doctoral training and research matters. More generally, the advent of federal research grants fostered an environment in which universities were more concerned about recruiting research faculty and making *them,* not the students, comfortable.

The seemingly absolute measures of competency available in mathematics made calculus or precalculus an attractive vehicle to many disciplines for screening prospective majors, even when calculus was needed by only the small fraction of their majors planning on doctoral study. Most mathematics departments maintained

comparatively high standards for performance and grading, leading to high failure rates for the students who were required to enroll. Some departments created special sections for meeting the needs of their client disciplines. In these sections, challenging physics-based problems were frequently eliminated, sometimes replaced by economics or biology examples. The continued lack of performance and high failure rates led departments to introduce placement examinations, which forced many students with three or four years of high school mathematics to repeat one or more high-school-level courses in college before enrolling in calculus or precalculus sections.

Three other developments should also be mentioned. Large section sizes and overworked faculty and TAs led to an elimination of grading of homework at many universities. Introduction-to-calculus courses in high school set the stage for superficial learning by skimming over the techniques of calculus with little understanding. After the 1960s, American TAs were increasingly replaced by foreign-born graduate students, whose language and cultural differences often provided a source for student complaints. All of these developments reinforced students' view of mathematics as an arbitrary hurdle.

By the 1980s, most faculty knew that the situation in calculus instruction was very unsatisfactory, but there were few efforts to do anything about it. The most successful was the applications-oriented calculus sequence integrating probability and discrete mathematics at Ohio Wesleyan in the mid 1970s, resulting in the text *Much Ado About Calculus* by Robert Wilson (Springer-Verlag). Supplementary materials such as the UMAP modules were of questionable value in courses where instructors were frequently struggling with students' deficiencies in algebra and were doomed to failure in courses where the grade was based completely on text-based mid-term and final exams. The prevailing sentiment was a mixture of guilt about being partners with students in allowing calculus to degenerate into superficial symbol-pushing (which still was hard for most students) and a natural human instinct to blame the situation on poorly prepared students—making it someone else's problem.

The Beginning of Change. The emphasis in the mathematics major on preparation for graduate study was turning employers and students off to mathematics. From 1970 to 1980, the number of mathematics majors dropped by 60%. The MAA 1981 curriculum report, *Recommendations for a General Mathematical Sciences Program,* reflected changes at many colleges which were attempting to "re-focus the mathematics major on the mathematical knowledge and modes of reasoning needed by the average mathematics student who is more likely to work in industry or teach in school than pursue a career based on advanced training in mathematics." While large numbers of students in the 1960s wanted to be mathematics majors and mathematics faculty could set high standards of performance, by 1980 the low interest in mathematics forced faculty to actively attract students—and not just highly talented students—to mathematics and to tailor their offerings more closely to the needs of these students. Instructional changes in response to these pressures both increased the number of students majoring in mathematics and set the stage for the calculus reform movement's focus on the needs of average students.

The development of discrete mathematics courses in the early 1980s was the final catalyst for action to rethink calculus instruction. Anthony Ralston obtained Sloan Foundation support to advocate that discrete mathematics become part of the beginning college mathematics curriculum, effectively replacing one semester in the standard three-semester calculus sequence. Ralston liked to be provocative. Instead of talking directly about discrete mathematics, he would have debates with people supporting the status quo in freshman mathematics instruction, namely, calculus. While calculus supporters recited the long list of scientific discoveries and modern technology based on calculus, Ralston challenged the effectiveness of current calculus to develop mathematical reasoning and problem solving skills. He also asserted that calculus had limited relevance to computer science.

The Discrete Mathematics Initiative of the Sloan Foundation was successful in publicizing discrete mathematics as well as developing new ideas and new texts for discrete mathematics courses. The plan of operation— a planning conference/workshop (which issued a well-publicized report) followed by multiple grants to develop curriculum and materials for new courses—was to be followed closely in the calculus reform movement.

Ron Douglas, who had taken the side of calculus several times in debates with Ralston, was motivated by these debates to initiate an effort to reform the teaching of calculus. The hurdle for the mathematics community was to stop feeling embarrassed about the sad state of calculus courses, to stop blaming high schools for underprepared students, and to admit openly that calculus instruction had serious problems that needed to be addressed. At the 1985 Joint Mathematics Meetings in Anaheim, California, Douglas, along with Steve Maurer of the Sloan Foundation, arranged a panel discussion "Calculus Instruction: Crucial but Ailing" to test the waters and see whether there were enough people who agreed that there was a problem in calculus. The panel drew an unexpectedly large audience, about 300 people, and the audience's reactions and questions revealed a broad agreement that calculus was indeed ailing and in need of revitalization.

Reacting primarily to concerns about mathematics education in the schools, the National Research Council (NRC) created the Mathematical Sciences Education Board (MSEB) to coordinate a broad-based effort involving school and higher education mathematicians, school superintendents, parent groups, and politicians to improve school mathematics. Like the changes in the mathematics major, MSEB's efforts focused on the needs of average students and other "clients" of school mathematics. MSEB also put heavy emphasis from the start on public relations and dissemination. The view was that educational change proceeded more through "evolution" than "immediate revolution" and that real, sustainable change had to be systemic in nature.

The Calculus Reform Movement

Launching the Reform Movement: the Tulane Conference and NSF Calculus Initiative

This section describes the start of the calculus reform movement with a mixture of documentary facts and an assessment of the early decisions that have proved valuable in helping the reform effort be so successful. The Tulane Conference provided an ideal start for calculus reform because it raised the right concerns and proposed effective strategies for accomplishing change. These strategies have been followed quite closely in most reform courses. Although the reform movement began with a conference of 25 people, they represented a broad constituency and reflected a grassroots set of values more closely associated with four-year colleges than doctoral institutions.

The Tulane Conference. With funding from the Sloan Foundation, Ron Douglas organized a working conference on revitalizing calculus instruction at Tulane University in conjunction with the January 1986 Joint Mathematical Meetings in New Orleans. The 25 participants at that conference included representatives from various client disciplines, as well as a diverse group of mathematicians, from two-year college faculty to research leaders. Mathematics faculty from four-year colleges formed the largest group. There were only four people who represented the research mathematics community. The proceedings of the conference were published as *Toward a Lean and Lively Calculus* (MAA Notes 6, Mathematical Association of America, 1986).

The optimistic and slightly elitist "can do" spirit of the post-World War II era was replaced by a pragmatic, inclusive, and cooperative spirit. As noted in the previous sections, this latter spirit reflected recent changes in the mathematics major and the plan of operation of MSEB, which was established in 1985. The consensus of workshop participants was that calculus was meeting neither the needs of its students nor those of client disciplines and subsequent mathematics courses. There were five problems with current calculus courses that participants kept mentioning:

- too few students successfully completed calculus;

- students were mindlessly implementing symbolic algorithms with no understanding and little facility at using calculus in subsequent mathematics courses;

- faculty were frustrated at the need to work so hard to help poorly prepared, poorly motivated students learn material that was a shadow of the calculus they had learned;

- calculus was being required as an unmotivated and unnecessary filter by some disciplines which made little use of it in their own courses; and

- mathematics was lagging other disciplines in the use of technology.

The conference had workshops on content, instructional methods, and implementation. From the content and methods workshops there evolved an agreement that greater conceptual understanding, developed through a variety of approaches using applications, numerical explorations with computers and calculators, and traditional algebraic methods, should be a guiding theme of the new calculus. For this to happen some topics would have to be omitted or de-emphasized. The content workshop suggestions—for example, de-emphasizing integration techniques and largely dropping related rates, l'Hospital's rule, and infinite series—have been adopted by many reform texts. The one area where the content report was mistaken was in its assumptions about technology. It urged calculator manufacturers to produce a moderately priced calculator that would find a root of a function and numerically integrate a function. It assumed graphics would require a PC, and for this reason it gave limited attention to graphical approaches.

The workshop on methods championed the notion, now widely accepted, that *how* calculus is taught is as important as *what* is taught. The methods report called

14

for activities to involve students, to make them active learners, and to "help them develop the ability to apply what they have learned with flexibility and resourcefulness." Cooperative learning, open-ended problems and projects, computer labs, and writing were all encouraged. It should be noted that the successes of Uri Treisman's PDP program, known to some workshop participants, had already shown the power of group learning.

Douglas recently said that, when he was thinking about the challenges facing calculus reform in planning the Tulane Conference, he felt that implementation would be the hardest aspect of reforming calculus because he was aware that there had been many individual mathematicians trying to change calculus courses for years and no effort had survived. The implementation workshop sketched a scenario in which new calculus courses would be independently developed at a carefully selected set of diverse institutions. Those efforts would use the Tulane Conference recommendations as a starting point.

After the conference, Douglas suggested to the implementation workshop group that success was more likely if there was initially a single large project with a writing team and a set of test sites monitored by a distinguished advisory committee, but the workshop report preferred more diversity. These two views of how to get reform off to a good start illustrate the many choices that had to be considered at the beginning of the reform effort.

The implementation report recommended a cooperative spirit in which all course developers get together to share ideas, and it called for careful evaluation techniques. It urged extensive public relations and an inclusive spirit of the reform effort, like MSEB's mode of operation, "aimed at making the new calculus the property of the entire mathematics community and not just a program of the participants of the New Orleans conference or for those involved in the initial projects." Looking at the bigger picture, the implementation report noted, "The build-up of Ph.D. programs caused a greater emphasis to be placed on research and led to a swing away from concern with teaching. A new calculus course can provide the impetus for a needed swing back toward more concern for teaching." The preceding quote was aimed at graduate students. While graduate students have not been involved as much as hoped, this quote did correctly foresee the impact of calculus reform on faculty.

In hindsight, the Tulane conference participants deserve great credit for proposing the following themes as

a blueprint for calculus reform:

A. Focusing on a conceptual understanding that uses a variety of intuitive graphical and numerical approaches and is geared to the needs of average students;

B. Emphasizing the importance of changing the modes of instruction and the use of technology to engage students as active learners; and

C. Fostering an inclusive spirit in the reform initiative and emphasizing the importance of cooperation and broad dissemination at every stage.

These three themes have turned out to be achievable. They have been adopted in virtually all reform efforts and have been broadly endorsed as appropriate and laudable components of the reform movement. They have been flexible enough to encompass a wide variety of efforts. The calculus reform movement has benefited tremendously from having unchanging, broadly accepted goals to guide it.

Theme A was anticipated by the 1950s Universal Mathematics project and the 1981 MAA mathematics major recommendations. More recently, dissatisfaction with existing calculus texts led the largest engineering society, the IEEE, to publish in 1985 a calculus text by Ash and Ash, *The Calculus Tutoring Book,* which "uses informal language and emphasizes geometric and physical reasoning." Theme C, also central to efforts of the MSEB, reflected the great concern about involving and energizing the mathematics community. Theme B, perhaps the most pervasive characteristic of resulting reform efforts, was largely new. Douglas's own experiences as mathematics chair at Stony Brook had convinced him that methods of instruction would prove to be critical, but he believed that this aspect of reform would garner little support from mathematicians. He wanted the issue developed at the conference because he felt that, once faculty got deeply engaged in calculus reform, they would be drawn into changing their teaching. This assessment has been completely validated.

Calculus reform efforts have benefited from a concurrent revolution in technology, especially the advent of reasonably priced graphing calculators and, to a lesser extent, relatively inexpensive personal computers. As an aside, we note that John Kemeny, a leader in developing the finite math side of *Universal Mathematics,* was already working in the mid-1950s to develop personal computing via terminals connected to a time-sharing computer and the language BASIC. A third of a century was also required before technological advances could give us an inexpensive, portable fulfillment of Kemeny's goal.

Initial Efforts to Start Reform Nationally and Within NSF. An NRC calculus panel headed by Douglas and an MAA calculus committee headed by Richard Anderson were created soon after the Tulane Conference to start planning for calculus reform. The MAA committee visited a dozen mathematics departments that were starting reform efforts. Neither committee initiated any reform activities in the year and a half following the Tulane conference, but faculty in many institutions were starting to plan calculus reform projects at the grassroots level, in expectation that the Sloan Foundation would fund a score or more reform efforts the way it funded discrete mathematics projects.

The reform movement might well have died away at this point had not the National Science Foundation started its Calculus Initiative. The NSF reinstituted the education directorate in 1987. Except for the graduate fellowship program and a small instrumentation program for four-year liberal arts colleges which began in 1985, there had been no education activity at the Foundation since 1981. The Neal Report, "Undergraduate Science, Mathematics and Engineering Education" (NSB 86–100), issued in March 1986, is credited with restarting undergraduate programs, the first being USEME (Undergraduate Science, Engineering, and Mathematics Education), which is now the Division of Undergraduate Education (DUE) in the Directorate of Education and Human Resources.

In summer 1986, NSF held separate workshops in several disciplines to plan a set of activities to implement the recommendations of the Neal Report. The workshop in mathematics included several Tulane Conference participants, and they successfully championed the cause of calculus reform as the natural choice for a mathematics initiative in a revived undergraduate education section in NSF. While this suggestion was implemented a year later, it was the Division of Mathematical Sciences on the research side of NSF, particularly then Division Director John Polking, Associate Director Judith Sunley, and program officer Louise Raphael (a Tulane workshop participant), who initially expressed interest and took leadership in supporting calculus reform, before the NSF un-

dergraduate education program was re-instituted. Ultimately, DMS and USEME jointly sponsored the calculus initiative, and Raphael was transferred from DMS to USEME to be the initial program director. When the program was expanded in 1992 to include the "Bridge to Calculus," the Division of Elementary, Secondary, and Informal Education formally joined as co-sponsors of the program. John Kenelly, John Bradley, William Haver, James H. Lightbourne, and others provided leadership for the NSF program in subsequent years.

To energize the mathematics community for the new NSF initiative, Douglas's NRC panel organized a national colloquium in October 1987, entitled "Calculus for a New Century," funded by the Sloan Foundation. Bernard Madison, then at the NRC, was the staff person assigned to Douglas's committee to carry out this monumental undertaking. More than 700 participants (at their own expense) attended the sessions to interact and express their opinions about changes in the teaching of calculus. It was the first time that the NRC auditorium was filled to capacity for a non-entertainment event. While most participants were hoping for simple solutions, the colloquium was not designed to go beyond the Tulane Conference's suggestions for how calculus should change, but rather sought to provide background information, to increase awareness and interest in the issues, and, foremost, to ratify the Tulane Conference recommendations for calculus reform. The conference proceedings were disseminated rapidly by the Mathematical Association of America (in early January, 1988) as MAA Notes 8, *Calculus for a New Century*. Six thousand copies have been sold.

The announcements of the new NSF Calculus Initiative were literally carried from the printer to the "Calculus for a New Century" conference. With interest in reform stimulated by the Tulane Workshop and NRC conference and with NSF support to nurture reform efforts, the calculus reform movement was under way.

The NSF Calculus Initiative. The National Science Foundation's Calculus Initiative began making awards in fall 1988. That a variety of different types efforts were supported, reflected the belief expressed in the June 1986 NSF workshop that there could be many viable approaches to reform. The NSF Calculus Initiative adopted the philosophy emerging from the Tulane Conference that modes of instruction and dissemination efforts were as important as the core content of the course.

In the first year of the NSF Calculus Initiative, much of the money went to 19 planning grants, with only five multi-year projects funded. The five multi-year projects reflected the diversity of approaches: one was to develop a new course and text, "Calculus in Context," with an applications-oriented approach; one was to develop a collection of open-ended student projects; one was for a cooperatively taught mathematics and physics course; one was technologically oriented, developing a user-friendly interface for students to use Maple to do calculus explorations; and one was to fund the creation of *UME Trends,* to disseminate information about calculus reform and other collegiate educational issues. The "Calculus in Context" project, the development of open-ended student projects, and *UME Trends* have worked out well in terms of substance and impact.

The initial idea for *UME Trends* came through a conversation between Bill LeVeque (then at the American Mathematical Society) and Lynn Steen following a presentation by Madison and Steen at the BMS Chairs' Colloquium. In responding to a question about publication of reports on calculus reform in traditional journals, Steen had commented that journal publications, as a rule, contained articles of "lasting interest"—those worth reading 10 to 20 years in the future. Educational reform articles seldom had that quality. Shortly after that, AMS prepared a proposal to NSF for the creation of a magazine for the quick publication of calculus reform articles. Thus with the leadership of AMS, the JPBM effort undertaking *UME Trends* was initiated.

In the second round of funding in 1989, six moderate-to-large projects for new courses, most with texts, were funded, along with 12 smaller projects aimed at local course reform. The text projects were the Harvard Consortium, Project CALC at Duke, Calculus and Mathematica at University of Illinois, the Oregon State project, and the Dartmouth Project. (Table 7, page 25, gives information about current adoptions of some of these text. The Dartmouth project does not have a text used elsewhere as of this date.) The small 1989 project by Keith Stroyan at the University of Iowa has grown into a larger project that also developed an innovative text. Also in 1989, a consortium of 26 midwest liberal arts colleges was funded to develop a set of books of resource materials for enriching calculus courses using traditional calculus texts. Over 6500 copies of the resulting five-volume set have been acquired by faculty.

In an early review of the NSF Calculus Initiative, Lida

K. Barrett and William Browder's assessment of proposals documents the attention being given to conceptual understanding, pedagogy, and technology. They wrote in *UME Trends,* October 1989:

> "In spite of the innovative nature of the proposals, the contemplated changes are not in the overall content of the course, but rather in the expectation of better student understanding; in greater stress on applications and links with other disciplines; in the utilization of numerical methods and computer techniques; and in encouraging a fresh approach to teaching."

Two more major course development projects with texts were funded in fall 1990 at St. Olaf College and Purdue University, along with 18 other smaller local projects for materials and integration of technology. Also, Frank Wattenberg separated from the "Calculus in Context" project to produce his own separate version of a reform text. In 1992, NSF funded a final major new text project by James Hurley at the University of Connecticut, aimed at a large network of high schools as well as college audiences. While Hurley had been developing computer materials to change calculus instruction since the mid-1980s, this project's text has only recently begun intensive development. The text-writing projects have all produced materials that have been used at numerous other institutions, although some texts have been used much more widely than others.

Starting in 1991, the NSF program shifted its focus from the development of pilot projects involving a limited number of calculus students to dissemination and large-scale implementation efforts, along with curriculum reform efforts aimed at pre- and post-calculus courses. The implementation involved high schools as well as colleges and universities. In fall 1992, about 200 institutions were using texts developed with NSF funding (some of the texts were not yet ready for pilot testing), while in fall 1994 over 800 colleges and universities and 300 high schools were using these texts. (Further information about text adoptions is given on page 25.)

As noted above, in recent years funding has been directed toward implementation and dissemination as well as extensions to multivariable calculus. Of the new awards made in FY 1992 and 1993, four directly addressed issues in precalculus, five were projects devoted to the development of programs in multivariable calculus or other courses beyond the first year calculus, but the

Table 1: NSF Awards by year

Year	Proposals	Awards	Funding (in millions)
FY 88	89	25	$2.3
FY 89	74	18	$4.6
FY 90	80	20	$2.4
FY 91	68	19	$2.4
FY 92	67	18	$4.6
FY 93	76	15	$3.2
FY 94	50	12	$2.7

majority (21) were broad dissemination or implementation projects. Those addressing courses beyond the first year calculus were "natural" extensions of existing projects but conducted at institutions different from the "originating project institution." The remaining three awards were for the development of a video tape on using "supercalculators" in curriculum reform and a series of calculus reform workshops directed by Don Small of the U. S. Military Academy and A. Wayne Roberts of Macalester College.

The NSF Calculus Initiative ended as a separate program after seven years in 1994. Its goals at the end were little changed, except for greater emphasis on dissemination and implementation. Since the inception of the program, regular meetings of the calculus project directors and other activities helped create the sense of joint efforts—a calculus reform movement—rather than just a collection of isolated projects. A chronological list of all awards made in the Calculus Program is contained in Appendix F. Table 1 summarizes the number of awards by year.

It should be noted that, over this period, more than a dozen Instrumentation and Laboratory Improvement (ILI) grants each year were also made to implement calculator- and computer-based laboratories for calculus courses. ILI applications by mathematicians increased substantially during the calculus reform movement. The calculus program directors sometimes referred Calculus Initiative proposals to ILI when their funds were expended. Program officer William Haver in USEME (later DUE) played a critical role in finding support for calculus projects from other NSF units. Mathematicians

in the ILI program and USEME programs in Faculty Enhancement and Teacher Preparation as well as EHR's Division of Research, Evaluation and Dissemination and Division of Elementary, Secondary, and Informal Education were all sympathetic to the calculus initiative and contributed funds to calculus-related projects.

The 127 awards made in the Calculus Initiative over the past seven years fall largely into seven categories: planning grants (20), primary curriculum development (23), technology-based calculus laboratories and software (15), projects/supplementary materials/cooperative learning (13), extensions to pre- and post-calculus courses (15), implementation (27), and dissemination/conferences (14).

It is worth noting that several reform projects have had a major impact on high school instruction in calculus, despite the lack of flexibility forced by the common AP Calculus examination. For example, the Dick/Patton text is currently used in over 250 high schools. NSF has funded a large effort to prepare AP calculus teachers to use graphing calculators so that modest technology-based reform would be possible at the high school level. This training was a major factor in the decision of the AP program to require graphing calculators in the AP examination beginning in spring 1995. In turn, the use of graphing calculators promotes and supports changes in the content of the AP calculus examination in the spirit of the calculus reform movement.

From the start, some critics of the calculus reform movement have argued that the changes being made were in fact merely cosmetic. A much deeper and more dramatic change in both curriculum and teaching would be needed if we were to correct the concerns that surfaced in the mathematics community during the mid-eighties. However, the NSF Calculus Initiative encouraged the calculus reform movement to be an inclusive, multi-faceted enterprise involving some quite radical efforts and modest changes as well as many middle-of-the-road efforts. As might be expected, the most successful reform text, by the Harvard Consortium, represents a middle-of-the-road reform. However, the modest and radical approaches have both played important roles. The modest approach makes reform more palatable to some; the radical approach allows for greater change that others find very much worth the extra effort, as well as providing useful ideas to middle-of-the-road efforts.

The Spread of Calculus Reform

Conditions Supporting Reform. Like the modernization of the undergraduate mathematics curriculum in the 1960s, the current calculus reform movement has benefited from a confluence of factors:

- A widespread dissatisfaction with the current state of calculus instruction, such as high failure rates, as detailed near the beginning of the previous section.

- The advent of technology that eliminated the need for some paper/pencil topics, such as graphing functions, and made accessible other topics, such as problems with integrals lacking closed-form solutions.

- The 1989 publication of the NCTM *Standards* and the broad support it received from academics and top elected officials, which reawakened a "can do" spirit and self-esteem in mathematics education that made calculus reform seem doable.

- With the end of the Cold War, federal funding of basic research in support of scientific competition with the Soviet Union reached a plateau as economic competitiveness and associated concerns about a well-educated workforce increased federal support for education. The 1986 NSF Neal Report had led to the reinstitution of the undergraduate education division, thus allowing collegiate mathematics to participate in NSF's growing educational mission.

- A declining number of 18-year-olds made improved retention important to sustain undergraduate enroll-

ments. This concern made funds available to mathematics departments that tried educational reforms to decrease failure rates in calculus.

- The creation of educationally oriented contributed paper sessions at MAA national and sectional meetings in 1987 which gave professional recognition to undergraduate mathematics innovation.

One very important difference between the 1960s collegiate changes and the present calculus reform is that the 1960s changes were led by the research community, as an initiative designed especially to produce more researchers. The current calculus reform movement has been more of a grass roots effort involving all sectors of the mathematics community (two-year colleges, four-year colleges, comprehensive universities, and doctoral universities) and directed towards improving calculus instruction for all students.

Along with general dissatisfaction with calculus instruction, there was also an absence of any guiding philosophy or plan for calculus instruction for a general college audience, like the techniques approach for engineers and the theoretical approach for math majors. When the Tulane Conference proposed conceptual understanding as a central theme, along with instructional strategies designed to engage the student more actively in mathematical reasoning, mathematics faculty interested in attempting change had a plan that both helped guide them and helped justify their efforts to colleagues and administrators. Finally, the proposed use of technology in calculus: (i) made calculus seem more contemporary to students; (ii) suggested the creation of a laboratory component to calculus courses which could foster mathematical exploration, projects, cooperative learning, and other pedagogical innovations; and (iii) provided the basis for many successful ILI equipment grants to mathematics departments. In short, the Tulane Conference's recommendations made it much easier to undertake calculus reform.

In 1989 the National Academy of Sciences published *Everybody Counts: A Report to the Nation on the Future of Mathematics Education,* and the National Council of Teachers of Mathematics released its *Curriculum and Evaluation Standards for School Mathematics.* These two documents were instrumental in focusing national attention on the need for change in mathematics education. Since then, the vision of school mathematics outlined in the NCTM *Standards* and the models for teaching suggested by NCTM's *Professional Standards for Teaching Mathematics* have been shaping major changes at the K–12 level. Calculus reform efforts are highly consistent with these emerging changes in school mathematics. Pedagogical strategies noted above for calculus are being used in schools to create classroom environments for realizing the vision of the *Standards*. Another point of coincidence in the two change efforts is the importance of posing problems that are interesting and meaningful to students.

Dissemination Efforts. An important factor in the success of the calculus reform movement was the extensive set of dissemination and information sharing activities that the NSF, the MAA, and the major projects sponsored. NSF calculus project directors have had an annual meeting to share ideas, as recommended by the Tulane implementation workshop. Many regional networks, several with NSF support, sprang up offering workshops and conferences on calculus reform. Three conferences sponsored by the Harvard Consortium project and their publisher, J. Wiley and Sons, have had a combined attendance of 1200. Calculus reform short courses, talks, and panel discussions have proliferated at national and regional MAA meetings and at locally sponsored events, drawing overflow audiences.

Harvard Consortium members have given about 50 short courses at professional meetings and locally sponsored events and another 100 invited talks and panel presentations. The Dick/Patton and Five Colleges teams have each given about 25 short courses and 50 talks on their projects. The other reform text developers have also given many short courses and talks. Collectively, the mathematicians associated with text projects have given about 150 short courses and over 300 talks about their projects. Besides the text project developers, many others involved in calculus reform have given talks and short courses about their, or others', materials.

The NSF funded workshops conducted by Don Small and Wayne Roberts have been quite successful in bringing mathematical sciences faculty an overview of the national calculus reform movement and a survey of the major text-producing calculus reform projects. The general format of these sessions has been not only for the workshop leaders to describe their approach to calculus reform, but also to address the other obstacles confronting curricular reform (e.g., need for new resources, skepti-

cism of client disciplines, resistance of colleagues). To date, about 20 such workshops for faculty have been conducted.

There have been several conferences about technology and mathematics, the largest being the annual International Conference on Technology in Collegiate Mathematics (ICTCM). Faculty participation in these ICTCM meetings was frequently cited in this study's in-depth survey of 62 institutions as a source of information that led to local change. The attendance at the annual conferences serves as an indicator of faculty interest and the growth of concerns about teaching using technology. The first ICTCM was held at Ohio State University in Columbus, Ohio, in 1988 with about 275 people attending. The next year's ICTCM attracted an attendance of 500. The third and fourth ICTCMs drew 800 people, and the two most recent ICTCMs had attendance in excess of 1500.

The NSF-supported *UME Trends,* which was initially sent free of charge to every mathematics department but is now self-supporting with close to 4000 subscriptions, reported on specific reform projects and themes and had thoughtful articles by leaders in the mathematics community discussing the need for change and the challenges to change. The MAA Subcommittee on Calculus Reform And the First Two Years (CRAFTY) sponsored talks, panels, and poster sessions at national meetings and sectional MAA meetings. The CRAFTY poster session in 1990 drew crowds of over 300 people (and still drew 300 at the 1994 annual joint meeting). Early in the reform movement in 1990, NSF funded CRAFTY to publish a lengthy report, *Priming the Pump for Calculus Reform,* MAA Notes 17, on some of the initial calculus reform projects. Over 6000 copies have been sold.

In 1987, the MAA initiated contributed paper sessions on educational topics at national meetings, allowing faculty engaged in calculus reform and other innovative educational efforts to get visible professional recognition for their efforts. While small at first, the number of education-related contributed-paper sessions grew quickly. At the 1994 annual joint AMS/MAA national meeting in Cincinnati, there were 280 education papers presented, some at AMS sponsored sessions, and 670 research papers. Moreover, their collective attendance levels approximately equal that at research sessions. Many ad hoc sessions to share information about calculus reform, with no set agenda, started appearing at national meetings and drew crowds overflowing into the corridors. Recent national meetings had sessions on reform in other introductory mathematics courses and a variety of pedagogical topics, including sessions on research in collegiate mathematics education. Over the past eight years, the national Joint Mathematics Meetings have gone from having two or three education-oriented events on the whole program to having dozens of educational events, almost all well attended. These activities have fostered an environment for questioning and reforming calculus and other collegiate mathematics instruction that may in the end prove to subsume and supersede the specific instructional changes of the calculus reform movement.

Several of the NSF-funded calculus reform projects involve two-year college faculty and students in substantive roles. Projects directed toward dissemination and implementation include two-year colleges as full participants. Major precalculus grants were awarded to the Maricopa Community College District (Arizona) and Suffolk Community College (New York). The American Mathematical Association of Two-Year Colleges (AMATYC) and its members have been active participants in the calculus reform efforts. The annual meetings of AMATYC also have sessions directed toward innovations in calculus and precalculus classrooms and the need for new pedagogical approaches. In fact, at the time of this writing, AMATYC is developing a set of curriculum and teaching standards for introductory level collegiate mathematics.

The in-depth survey indicated that at virtually every one of the mathematics departments where some level of reform, however modest, was occurring, the faculty collectively had attended many presentations, often a dozen or more, about calculus reform at conferences, professional meetings, or workshops.

The Faculty Who Implemented Reform

As the reform movements developed, there was an evolution in the types of mathematicians who played key roles. The 1986 Tulane Conference was organized by Ron Douglas. The Washington Conference, "Calculus for a New Century," was organized by Douglas's NRC calculus panel. The NSF research side provided some of the funds for the Calculus Initiative. However, the NSF Calculus Initiative and MAA calculus efforts nurtured a broad-based grass roots structure in the reform movement. The NSF-supported projects involved some people who viewed themselves as primarily associated with the American Mathematical Society and who had

recently held NSF research grants, but most funded projects were undertaken by people who were primarily associated with the Mathematical Association of America and who had a previous involvement in undergraduate mathematics education. The MAA's CRAFTY and Committee on the Undergraduate Program in Mathematics (CUPM) became the committees most closely associated with reform calculus. The energy and networking skills of Lynn Steen as chair of CUPM had a very positive impact on undergraduate mathematics education generally and its visibility among professional societies and federal agencies.

Most reform texts grew out of NSF-funded projects at doctoral institutions even though a minority of NSF calculus grants went to doctoral institutions. The faculty in text projects generally were already active in educational matters, and few had NSF research grants. The dominance of doctoral institutions in text-writing projects is probably attributable to their lighter teaching loads, which gave faculty more time for large undertakings, and to the experience of faculty at doctoral institutions in undertaking large projects. The latter explanation is supported by the fact that the two text projects at non-doctoral institutions were at St. Olaf College and Five Colleges, whose mathematics departments had past success in securing large grants for educational projects.

It is surprising that the level of reform at doctoral, master's and bachelor's degree institutions in the study's 1992 and 1994 surveys were fairly similar. Due to factors discussed in the next chapter, two-year colleges appear to lag a bit in implementing reform. Interestingly, master's degree (comprehensive) universities were less active than four-year colleges and doctoral-granting universities in early large-scale reform efforts. The spring 1992 survey of 745 institutions indicated that only 6% of responding master's degree universities had major reform efforts under way, as opposed to 13% of responding doctoral universities and 15% of responding four-year colleges. A later survey in spring 1994 indicated an increase in major reform at master's universities to 24%, slightly higher than at doctoral institutions.

Lower-Profile Reform Efforts. While attention has come to be focused on the larger projects that produced texts for wide dissemination, a recent survey by this assessment study indicated that about half of the reform efforts are using traditional texts supplemented with materials and activities developed locally or nationally. These materials and activities include more applications, use of technology, student projects, cooperative learning groups, and student writing. The nationally publicized reform effort encouraged individuals who had long been thinking about ways to improve calculus instruction and created a climate that allowed concerned individuals to get permission from department chairs to develop experimental sections.

Many NSF-funded projects were small efforts seeking to produce changes in instruction at only one institution or a small group of institutions. Technology was frequently a critical part of the project, whether funded by the Calculus program, the ILI program, private foundations, or local sources. These projects followed the original recommendations of the Tulane Conference quite closely. They stressed conceptual learning and active involvement of students in the learning process, using technology, and a mixture of student projects, cooperative learning, and extensive writing. Several projects produced supplementary materials that were published for use by others. By spring 1991, 28 books of supplementary materials for calculus were already in print (see the article by A. Ostebee in *UME Trends,* August 1991). At this writing, the number is estimated to be well over 50. The following data is one indicator of the large market for such materials. After 2400 copies of the NSF-sponsored five-volume set of supplementary materials, *Resources for Calculus,* were distributed one to each mathematics department, an additional 4000 copies of the set have been purchased by individuals through the MAA. The New Mexico State University project's *Student Research Projects in Calculus* has sold 5600 copies. These materials are frequently used as resources for high school calculus courses as well as in collegiate classrooms.

The Role of Research Mathematicians in Reform. Starting from the Tulane conference, there had been a great concern about getting research mathematicians involved in the reform efforts. An article by Peter Lax in *UME Trends,* May 1990, urged they become more involved in educational reform and was critical about their role in the high failure rates in university calculus courses: "I have little doubt that the culprits are the vast majority of research mathematicians." Not surprisingly, faculty with summer research funding proved uninterested or unwilling to set aside (or greatly reduce) their funded research programs to undertake calculus reform projects.

However, some mathematicians with research funding did undertake unfunded reform efforts. Most notable is the reform calculus course at the University of Washington using a course manuscript developed by Neal Koblitz and colleagues. This effort slightly pre-dates better known efforts such as the Harvard Consortium's and was developed locally with no NSF support and little interaction with national calculus reform efforts. It uses calculators, computers, and cooperative learning and emphasizes conceptual understanding and applied reasoning skills. It decreases time spent on integration techniques, dropped l'Hospital's Rule and some other topics and increased tangent line approximations and differential equations. In short, the University of Washington independently evolved the same type of reformed calculus as the major NSF projects.

Except for a few faculty at Harvard, Stanford, and Illinois, research mathematicians at leading (top 25) departments did not become involved in calculus reform until the Calculus Initiative's focus turned to large-scale implementation, although the University of Illinois had a major Mathematica-based project involving a text used in many calculus sections and Cornell had a small project. Reform has come to these departments in the form of the adoption of a reform text, typically the middle-of-the-road Harvard Consortium text. The initial adoptions were largely instigated by department chairs who had become interested in undergraduate education. The first was Stony Brook, which in fall 1992 adopted a reform text for all sections of its engineering/science calculus sequence. Don Lewis, chair at the University of Michigan, argued forcefully to research mathematicians at his home institution and to the general research community about the importance of improving undergraduate education. Thus, the Michigan mathematics department adopted a reform text coursewide in fall 1994, following two years of smaller-scale preparatory efforts, as part of a larger re-vitalization of its undergraduate mathematics program. A favorite line of Lewis's about priorities for research mathematicians was, "An NSF grant providing two months of summer salary is icing on the cake. The cake is calculus." In 1993 and 1994, the AMS-oriented Mathematics and Educational Reform network (MER) held three-day conferences (funded by NSF) about calculus efforts. By spring 1994, over 20% of doctoral-granting mathematics departments reported a major reform in calculus under way, and the number seems to be growing rapidly in the wake of relatively trouble-free implementation experiences at leading universities such as Stanford and Michigan.

Changes Produced by the Calculus Reform Movement

Amount of Reform Activities

Surveys Conducted by this Study. This assessment study undertook two short surveys in spring 1992 and spring 1994 which generated 745 and 1067 responses, respectively. The study also mailed out in-depth surveys in spring 1993 and spring 1994 to a subset of mathematics departments that had indicated in one of the short surveys that some level of calculus reform was under way in their institution. (Appendix B includes copies of these survey forms.) The number of institutions selected from each of the levels (doctoral, master's, bachelor's, associate) were in proportion to the number of responses to the short survey received from that level. The selection of schools was done in random fashion, with some minor, subsequent adjustments to ensure geographic diversity within levels. The in-depth surveys received 62 responses (18 in 1993 and 44 in 1994), involving 11 doctoral institutions, 13 master's degree institutions, 26 bachelor's degree institutions, and 12 two-year colleges.

It is interesting to note that the response rate to both surveys was positively correlated with the degree offered—the higher the degree, the higher the response rate. The response rate from doctoral institutions in the second survey was very high, 137 out of a total of the 165 United States doctoral institutions, or 83% (the number 150 of doctoral responses in Table 3 includes 13 cases of institutions with two responses from different mathematical sciences departments).

The in-depth survey indicated that there had been few changes in the credit hour structure of the calculus courses offered at the schools and that the predominant—"mainstream"—calculus course is still a three-semester sequence for mathematics/physical science/engineering majors, although at many institutions such targeted students make up a minority of the enrollment of this course. The 1990 CBMS survey estimated that in fall 1990, about 360,000 students were enrolled in the first year of the mainstream course and about 200,000 in other versions of first-year calculus. The first semester of the mainstream course covers limits, continuity, differentiation, and introduction to the integral. The second semester covers methods and applications of integration and series. The third semester covers vectors and vector-valued functions, partial differentiation, and multiple integration. About half of the respondents to the in-depth survey indicated that they offered other versions of calculus for business, social science, and natural science majors.

What Constitutes Reform. Just how much innovation is necessary to have a calculus course qualify as being a "reformed" course that produces real change? If one uses graphing calculators simply to graph the functions in the course or if one uses a computer algebra system to do the standard textbook problems, would that constitute a "reformed" course? To many, the answer would be no. However, responses to this study's in-depth survey indicate that meaningful changes in pedagogy and student attitudes have occurred at institutions adopting such quite modest reform efforts.

In this assessment study, the term "modest reform" will refer to situations where some change in instruction has occurred (use of technology, establishment of learning groups, etc.), and either a reform text has been adopted or supplemental materials have been developed for use with a traditional text. "Major reform" refers to situations where the department has adopted a reform text in most or all calculus sections and includes changed instructional approaches.

Calculus reform in one sense is trying to dispel the criticism that the typical calculus homework assignment of 25 similar exercises sends students two unfortunate messages about the nature of mathematics:

a) If you know what you should know about mathematics, any problem can be done in 10 minutes;

b) If you encounter a problem in mathematics that you can't solve, reread the previous five pages of text and look at the examples.

Calculus reformers who are impatiently trying to get "real change"— dramatic revision in both content and pedagogy—see most faculty as being unwilling to make the necessary profound changes in their own beliefs and then devote the time and energy that it would take to achieve real and lasting change. Sometimes where institutions have been adopting or adapting projects developed elsewhere, the changes are only "cosmetic" and no

Table 2: 1992 Short Survey: Response by Institution Degree Type and Level of Change
(Percentages in parentheses are with respect to numbers in each row)

Highest degree /Change	None	Modest	Major	Total Responses	Total Institutions
Doctoral degree	34 (40%)	39 (47%)	11 (13%)	84	165
Master's degree	45 (38%)	65 (56%)	7 (6%)	117	236
Bachelor's degree	133 (40%)	151 (45%)	51 (15%)	335	1020
Associate degree	115 (55%)	79 (38%)	15 (7%)	209	1018
BS + MS + PhD	212 (39%)	255 (48%)	69 (13%)	536	1421
Total	327 (44%)	334 (45%)	84 (11%)	745	2439

Table 3: 1994 Short Survey: Response by Institution Degree Type and Level of Change
(Percentages in parentheses are with respect to numbers in each row.)

Highest degree /Change	None	Modest	Major	Total Responses	Total Institutions
Doctoral degree	36 (24%)	82 (55%)	32 (21%)	150	165
Master's degree	35 (26%)	67 (50%)	33 (24%)	135	236
Bachelor's degree	116 (26%)	207 (47%)	120 (27%)	443	1020
Associate degree	147 (46%)	126 (39%)	47 (15%)	320	1018
BS + MS + PhD	187 (26%)	356 (49%)	185 (25%)	728	1421
Total	334 (32%)	482 (46%)	232 (22%)	1048	2439

(NOTE: Of the 1067 responses to the 1994 short survey, 19 did not respond to this question.)

real change in the learning of students is taking place. According to these critics, one has a "reformed calculus" effort while very little has really changed in the classroom.

Levels of Reform Determined from Short Surveys. Tables 2 and 3 summarize the information obtained from the two short surveys about the level of change by type of institution in 1992 and 1994. Table 2 shows that calculus reform was already quite active in spring 1992 before the implementation efforts for reform texts, although major reform was occurring at only 11% of responding institutions. Table 3 shows that by spring 1994 three-quarters of responding four-year colleges and universities had some sort of reform effort under way and almost a quarter had implemented a major reform effort. The tables show a percentage increase in institutions with "some change" and institutions with "major change" at every level of in-

stitution from 1992 to 1994.

One might expect that institutions involved in calculus reform would respond at a higher rate in the surveys than institutions not so involved. One measure of the relative responsiveness of institutions to reform is the statistical method of capture-recapture which looks at the percentage of respondents to the first survey who also responded to the second survey. There were 470 institutions responding to both surveys. That is, 63% of the first-survey respondents answered the second survey. This compares with the percentage of 43% of all higher education institutions that responded to the second survey. An analysis of the institutions that responded to both surveys shows that their level of reform was distributed approximately the same as that for all respondents to the second survey. Thus, we argue that the second survey, and by symmetry the first survey, are not significantly biased by over-response from reform-oriented institutions.

Table 4: 1992 Survey: Indication of Breadth of Change by Degree Type Among the 394 Institutions Reporting Change

Institution/Sections	Experimental	Course-wide
Doctoral	44	12
Master's	54	15
Bachelor's	78	120
Associate	30	41
Total	206 (52%)	188 (48%)

Table 5: 1994 Survey: Indication of Level of Change by Breadth of Change and Degree Type Among the 733 Institutions Reporting Change

School/Sections	Modest Exper	Major Exper	Total Exper	Modest All	Major All	Total All
Doctoral	62	10	72 (59%)	26	24	50 (41%)
Master's	43	10	53 (50%)	27	26	53 (50%)
Bachelor's	66	22	88 (39%)	138	102	140 (61%)
Associate	50	6	56 (32%)	79	42	121 (68%)
Total (733)	221	48	269 (37%)	270	194	464 (63%)

(Appendix C contains further details of this analysis.)

A comparison of the level of change in 1992 and 1994 among institutions responding to both surveys indicated that at each of the four degree levels (doctoral, master's, bachelor's, associate), the pattern is one of movement toward reform, but with the majority of schools remaining within the same category of change. For the four levels, 59%, 67%, 56% and 57%, respectively, report no change in reform activity over the two year period. However, 33%, 36%, 32%, and 28%, respectively, report an increase of at least one level of reform activity over that same time period. More data about the institutions that responded to both short surveys is given in Appendix C.

While some two-year colleges are reform leaders in their regions, as a group, two-year colleges showed the lowest level of change at both times. However, the rate of growth in two-year institutions indicating some change (from 45% in 1992 to 54% in 1994) is about the same as the growth rate at other types of institutions (from 61% to 74%). Heavy teaching loads, reliance on part-time faculty to do a large amount of teaching, and dependence on matching calculus course content to that at colleges and universities with which they articulate, seem natural explanations for this lag at two-year colleges.

As noted earlier, there has been a large increase in the amount of major change at master's degree (comprehensive) institutions. Only 6% of such responding institutions reported major reform efforts under way in 1992, less than half the percentage of responding doctoral universities (13%) and responding four-year colleges (15%). In the 1994 survey, major reform increased to 24% of responding master's institutions, slightly higher than at doctoral institutions.

Tables 4 and 5 summarize information about the number of calculus sections involved in change among institutions indicating some level of change. (The 1994 survey requested more detailed information.) Table 4 shows that doctoral and Master's institutions initially focused reform efforts on experimental sections, while reform in all sections predominated at four-year colleges and efforts were split between experimental sections and all sections at two-year colleges. The latter two types of institutions frequently have modest calculus enrollments and just one calculus course.

In Table 5, the 194 institutions reporting major change in all sections represent 18% of all respondents in the 1994 survey. The comparable percentage in the 1992 survey was 5.5% (not shown in Table 4). This represents

Table 6: Spring 1994 In-Depth Survey: Future Reform Plans Among 44 Institutions

School/Change	All Sections Already Reformed	Soon All Reformed	Soon Some Increase	No Change Planned	Total
Doctoral	1	3	2	1	7
Master's	0	6	1	1	8
Bachelor's	8	5	3	4	20
Associate	4	4	1	0	9

a very substantial increase over two years for total reform for the schools responding. It is worth noting that 30 schools in the 1994 survey indicated the use of reform in both experimental sections and all sections. Such responses were interpreted to mean that all sections are using some reform but a few sections are experimenting with a more radical version of reform. The in-depth surveys indicated that experimental sections with two or more different reform texts have not been uncommon.

The number of students currently enrolled in reform calculus sections at the 600 institutions responding to the request for student data in the spring 1994 survey was approximately 125,000. This provides a very conservative lower bound on reform enrollments in spring 1994. Extrapolating from this number, a more reasonable estimate for all institutions (those responding to the survey but not giving a numerical response to this question and those not responding) would be around 175,000 students in spring 1994. A cross-check on this number is that the 1994 in-depth survey indicated that slightly under half of reform efforts use reform texts and Table 7 below suggests that about 82,000 reformed texts were in use during spring 1994 (82,000 is 47% of 175,000). However, the number 175,000 is clearly quite soft and could be criticized on many grounds. A more conservative statement would be that at least 150,000 students were enrolled in reform courses in spring 1994.

The CBMS 1990 enrollment survey (see Appendix D) estimated that approximately 560,000 students were taking first-year calculus in fall 1990, and more recent, but less accurate, AMS data indicate this number has not changed during the 1990's. However, calculus enrollments are about 20% lower in the spring because many students take only one semester of calculus beginning in the fall, and so 470,000 is a reasonable estimate for the total spring 1994 first-year calculus enrollments. Hence about 32% (150,000/470,000) of spring 1994 calculus enrollments were in reform calculus courses.

The number of faculty currently involved in reform calculus teaching at the 758 institutions responding to the request for faculty data in the spring 1994 survey was approximately 3300 (about 4.5 faculty per institution). There were 1756 faculty involved at institutions reporting modest change and 1393 faculty at institutions reporting major change. A reasonable estimate for faculty involved in reform efforts at all institutions would be about 4500. There is a total of about 26,000 mathematics faculty in institutions of higher education.

Survey Results about Future Changes. Table 6 shows data from the in-depth 1994 survey about future reform plans. Note that among the 31 departments which had not yet adopted a reformed approach in all sections, 18 (58%) expected course-wide adoption of reform materials within two years for all calculus classes (three of the 18 actually indicated adoption for all mainstream classes but not business calculus classes). Another measure of the growth in course-wide use of reform materials is the following data about the use of the Ostebee/Zorn text: when their preliminary text first appeared in fall 1992, one quarter of the 28 institutional adoptions used the text course-wide; in fall 1994 (still preliminary edition), over half of the 50 new institutional adoptions were for course-wide use.

Reform Texts. In the spring 1994 in-depth survey, 40% of institutions indicating that some level of reform was under way said that they were using a reform text. Most were using supplementary materials, usually in published form. Nonetheless, calculus reform is increasingly becoming associated with one of the reform texts. Projections based on Table 6 suggest that the majority of reform efforts that are under way in 1994–95 are using reform texts. Table 7 summarizes recent data on 1993–94 and 1994–95 adoptions and sales of reform texts. Note that regular (non-preliminary) editions of four reform texts have become available only in the past year, while the rest are still in preliminary form.

Table 7: Survey of Publishers and Authors on Use of Reform Texts

Texts	Preliminary /Regular Edition	Institutions Adopting 1993–94	Adoptions 1994–95	Sales 1993–94	Sales Fall 94
Harvard Consortium	Reg	250*	380*	50,000	60,000
Ostebee/Zorn	Prel	70	116	9,000	15,000
Dick/Patton	Prel	45*	50*	10,400	12,500
Project CALC (Duke)	Prel	40	45	5,500	6,500
Five Colleges (Calculus in Context)	Prel	21	35	3,000	4,000
Calculus with Mathematica	Reg	16	20	3,000	4,000
Stroyan (Iowa)	Reg	10	15	2,000	3,000
Wattenberg	Reg	10	12	2,000	2,500
C^4L/Purdue	Prel	35	40	1,200	1,200
Totals		552	613	86,100	108,700

* Additionally, Harvard Consortium is used in about 125 high schools and Dick/Patton is used in about 250 high schools.

At present, the Harvard Consortium text clearly is the dominant reform text. This preference is heightened at doctoral institutions, where over 80% of the time it is the reform text of choice. There is slower, but still steady, growth of the reform texts with heavier use of technology. For example, the University of Illinois has more and more sections using Calculus with Mathematica. Last year, the Illinois biology department paid for additional math TAs to help convert the calculus for biologists' classes over to Calculus with Mathematica. There are increasing numbers of text reference materials being developed with Mathematica as a basis. The materials developed by Keith Stroyan are in this arena.

Despite fears of a backlash at institutions where reform would be dropped after coming too quickly or not being properly implemented, the reform texts have seen a very low drop rate. Of the 110 institutions using the Harvard Consortium text in the 1992–93 academic year, only three did not continue using it the next year. Of the 315 institutions using it in 1993–94, only 15 (5%) did not continue. Dick/Patton, Ostebee/Zorn, and Five Colleges have also had drop rates around 5%. In some cases of a text being dropped, the reason was a shift to another reform text. In other cases the text was dropped because there was one experimental section taught by a professor who was going on leave. A minor number of the drop cases involved a decision to drop all reform efforts for the foreseeable future. In sum, the number of cases of backlash to reform texts to date has been extremely low, around 2%.

Content Changes

As stated repeatedly, calculus reform has come to focus more on how calculus is taught and learned and less on what is actually taught. The responses to questions about content in this study's in-depth surveys frequently talked about how learning is occurring rather than what is learned. The overall focus on raising students' conceptual understanding, problem-solving skills, analytic ability, and transferability of calculus skills to work in other disciplines has led to general changes not in the list of topics and techniques covered, but in how these topics are developed. Some content changes have also been driven by what use of technology occurs. One respondent indicated that this restructuring of the knowledge-based objectives for the course might be viewed as changing the skyline from a series of homogeneous-looking buildings to a skyline of a few skyscrapers with several adjoining short buildings.

The data in this section and the following sections are drawn from a variety of sources: the in-depth surveys conducted by this study, a survey done by a doctoral student of M. Kathleen Heid at Pennsylvania State University, presentations of reform projects at various meetings, and interviews with individuals implementing reform.

There is a substantial range in the level of theory in reform texts, with Ostebee/Zorn high on theory and the Harvard Consortium quite low. Several of the texts that depend heavily on technology are also fairly rigorous in their use of definitions and theorems, although, in some, few theorems are formally proved.

What is Added.   In modest reform courses, where calculator/computer activities in problem sessions complement lectures from a traditional text, there is little change in the content of the lectures. The technology-based activities seek to give graphical and numerical experiences to enhance understanding of calculus concepts. Activities to develop mathematical reasoning usually involve applied models. Almost all institutions with modest reforms and some with major reforms indicated in responding to the project's survey that they were teaching all the material that they had taught their students before and that their course syllabi had not been changed. These responses indicate that there may be cause for concern that courses, instead of getting "leaner," may become "packed" in the name of calculus reform.

In search of greater conceptual understanding of the basic operations of calculus, all reform efforts give increased attention to tangent line approximations in differentiation and to Riemann sums in integration. They use numerical methods to approximate the derivative and the definite integral.

All reform texts give greater attention to functions, going beyond simply their algebraic properties. They ask: how do different types of physical processes give rise to different functional models? Functions' mathematical properties, in algebraic, numerical and graphical terms, are closely tied to properties of these processes. As mathematical analysis is applied to these functions, one interprets the results in algebraic, numerical, and visual terms and in terms of the underlying processes. Frequently reform texts will have discussions about functions that are defined only visually, as a graph (with no algebraic formula). Functions to be integrated are sometimes given in tabular form, e.g., find the approximate volume of a tree from the ground up to 10 feet if at one foot above ground the tree has diameter 10, at two feet above ground diameter 8, etc. Reform texts go more deeply into the inter-

pretation and analysis of applied models than was the norm in traditional calculus texts. Most give greater attention to Taylor series. Most have a chapter on simple differential equations and their applications.

What is Removed.   It is a mistake to put much weight on what reform courses are omitting, but, with increased attention to the preceding issues and topics, something has to be decreased. Virtually all reform courses have streamlined the discussion of techniques of integration, and trigonometric substitutions have been almost eliminated in some courses. Generally reform courses have downplayed symbolic computation skills. Graphing calculators have largely eliminated curve sketching techniques in most reform courses. Related rate problems are often missing. Limit theorems and calculus theorems such as the mean value and intermediate value theorem are also downplayed. For example, the Harvard Consortium text mentions the mean value theorem quickly in passing while developing error bounds in Taylor series. Some aspects of infinite series are gone in many reform courses. Note that this choice of de-emphasized topics follows closely the recommendations from the Tulane Conference. And this is not surprising given that the text writing teams from both Harvard and St. Olaf had members who served on the Tulane Conference Content Committee.

Overview of Reform Texts.   Reform texts are naturally classified into two categories: moderate use of technology and heavy use of technology. The latter usually are linked to Mathematica or MAPLE, although ISETL (at Purdue) and DERIVE (at Duke) are also used. The texts linked to computer algebra systems generally require substantial exploration and discovery through computer activities, which take more time and deeply engage the student in the learning process. Almost all reform texts discuss topics that fall outside the traditional calculus syllabus. Their opening chapters exemplify this larger agenda. All reform texts begin with a chapter or more on some mixture of (i) mathematical modeling and (ii) uses and properties of functions from a graphical and applications point of view. The modeling often involves difference equations. All texts present differentiation before integration. Differentiation is presented as a limiting case in rate-of-change questions that are usually motivated by applied settings. Some applications of differentiation are often presented before developing the formulas for derivatives of elementary functions. In a few texts, some simple differential equations are discussed at this stage.

Table 8: Spring 1993 and 1994 In-Depth Survey: Use of Pedagogical Activities Among 62 Institutions

Activity	Major	Moderate	Little	None
Calculators	36	13	0	13
Computers	10	14	11	27
Cooperative Learning	6	20	21	15
Student Projects	6	23	13	20
Extensive Writing	4	18	18	22
Modeling Applications	20	20	7	15

Integration is begun with graphical and applied situations that motivate the integral. There is extensive numerical exploration of limiting left- and right-hand sums. Techniques of integration are downplayed. There is usually an extended discussion of numerical methods including informal discussions of their convergence rates. There is often "applied" epsilon-delta reasoning: in the numerical approximation of a particular definite integral of a convex function, how large must n, the number of intervals, be so that the difference between estimates given by the trapezoidal rule (bounding from above) and the midpoint rule (bounding from below) is less than 10^k for various values of k? Applications go into greater depth and technicalities than in traditional texts. The texts use real (messy) data, e.g., eight-digit physical constants. Following the extensive applications of the integral, there is sometimes a chapter on differential equations. The texts usually finish with Taylor series.

Pedagogy Changes

In almost all the NSF-funded projects, whether local efforts or those producing texts, new instructional methods are introduced that reduce tedious calculations and try to involve students more directly. Activities are designed to make students active learners with more self-confidence about their understanding of calculus concepts and their mathematical reasoning skill. Pedagogical approaches include more in-depth and open-ended problems, exploration with technology, student projects, cooperative learning, writing, and the use of multiple representations of situations—numerical, graphical, and algebraic.

David Smith of Duke University uses "coaching" as a metaphor for what many involved in calculus reform are doing in the classroom. He points out that students understand that no one learns to play basketball or the

violin by listening to lectures about dribbling or bowing. Those students also understand that people who lack the talent to be a varsity athlete or a concertmaster can nevertheless learn to play well enough to achieve some personal satisfaction. But to do so, they must start playing—preferably under the guidance of a good coach. (Note that violin "coaches" are usually called "teachers.") Students also recognize that athletic and artistic endeavors are more satisfying when practiced with others and not done in isolation. Consequently, group work in the mathematics classroom is consistent with the coaching metaphor.

This study's in-depth survey elicited Table 8 of responses about different pedagogical approaches. Note that the levels of activity were sometimes reported in narrative form and had to be categorized as Major, Moderate, Little, or None by project staff. Also, the difference between Major and Moderate is necessarily quite subjective.

This data contains perhaps the most impressive information about the impact of reform in the whole in-depth survey. What is surprising is the amount of change in pedagogy other than the use of technology, given the established teaching habits of faculty and the effort required to change those habits. Since the majority of reform projects were still using supplementary materials in conjunction with a traditional calculus text at the time of the survey, this data indicates that these so-called modest reform projects are doing more than just "playing around" with calculators and computers. Moreover, the reports below about changing student attitudes in modest reform efforts give added support to the notion that modest efforts may be accomplishing more reform than their critics realize.

Use of Technology. The 80% of respondents in Table 8 saying that calculators are used in a substantial way is expected and validates both the word-of-mouth consensus about heavy calculator usage in reform and the large number of technology-oriented supplementary materials that have been published. Both students and faculty find graphing calculators an appealing tool. Essentially all reform efforts appear to use either calculators or computers. In fact, the use of graphing calculators is so widespread that many people no longer consider their use as a sign of calculus reform. With graphing calculators being required on the AP Calculus exam beginning in the spring 1995, incoming students will be insisting that they be used in all college calculus courses. Tom Tucker, of Colgate University and initial chair of CRAFTY, has asserted that meaningful reform can take place with very little technology, but his point now seems moot.

Survey responses about reform courses using traditional texts said that they were using technology-based supplements. A few institutions mentioned using the calculator-expanded version of the Thomas/Finney calculus text, whose authors are Thomas, Finney, Demana, and Waits, for most calculus sections while using a reform text or extensive supplementary materials in a few sections.

The survey revealed some information about the types of calculators being used, but technology is changing fast enough to make this data quickly obsolete. It is worth noting that a few institutions were using powerful HP calculators. One interesting finding was that some institutions do not require graphing calculators to be purchased, but instead have them in labs or rent them out to students. This was most likely to occur at two-year colleges and public four-year colleges.

The use of computer algebra software (CAS) to support calculus reform was widespread among institutions reporting computer usage. Here there were clear differences in the type of school reporting computer usage. While the AA, BS, and MS institutions reported solid usage, the PhD-level institutions tended to defer such usage to the Calculus III level and higher. This may be a direct result of the numbers of students involved and the availability of adequate resources to carry out the student work at the computers (or terminals). Among the schools identifying a CAS system they used, the most frequently mentioned system was DERIVE, followed by MATHEMATICA. Several reported using locally developed software. Also, several respondents reported using more than one product.

Several reformers involved in technology-based projects, such as Ed Dubinsky and Jerry Uhl, have complained that there is too much emphasis on technology as an end in itself. They argue that technology should be only a tool to assist students in learning. The C⁴L project of Dubinsky, Mathews, and Schwingendorf uses programming in ISETL or MAPLE as a means to stimulate students to undertake needed mental constructions.

As noted below in the section on student attitudes, the use of technology has played a major role in creating a very positive image for mathematics: many students are coming to associate the use of technology with mathematics. There are also practical dividends with technology. For example, Arizona has disks with a tutorial to prepare for calculus along with practice calculus readiness tests that are sent to all entering freshmen during the summer.

Responses to the project's surveys revealed that acquiring computers has been a major impediment to reform at many institutions. It is thus reasonable to speculate that greater use of computers would be occurring if more departments could get adequate funding to purchase them.

Cooperative Learning.   In Table 8, 42% of respondents made substantial use of cooperative learning while another 34% used the method, but infrequently. Typically this group learning occurs in problem or laboratory sections while working on projects or in-depth problem sets. A couple of institutions reported using a Treisman-type approach of running problem sessions in small groups, although these institutions had comparatively small calculus enrollments. One of the major reform projects, the Purdue C⁴L effort of Dubinsky, Mathews, and Schwingendorf, which is being implemented at 40 institutions, places a very heavy emphasis on group learning, in part because they believe that explaining an idea to another student forces a person to think more deeply about it.

Student Projects.   Almost half of the respondents in Table 8 reported substantial use of projects. The number was lower at doctoral institutions; only three out of 11 reported substantial use of projects. Several respondents indicated, either directly or indirectly, that their project use was more the consideration of special problems than a lengthier consideration of an application requiring extended student work. One school indicated that their use of projects was strictly for extra credit.

Writing.   In Table 8, 35% of respondents reported substantial use of writing. Most respondents using substantial writing indicated that this occurred through write-ups of projects and in-depth problems. One institution indicated that it used a commercially published book specifically aimed at uniting writing with project work, and one indicated that it was requiring students to maintain journals. The remainder reporting a substantial use indicated simply "an increased emphasis" on writing. Only two of the 11 doctoral institutions indicated substantial use of writing. This is probably because their large calculus enrollments make grading of written work on a regular basis more difficult.

Modeling and Applications.   In Table 8, 65% of respondents reported substantial use of applications and applied modeling. Several made special mention of their use of modeling to provide a conceptual strengthening to the calculus curriculum while attempting to motivate the material for their students. Because some applications were already in many traditional courses, some respondents may have indicated little applied work when they were actually meaning little increase in applications. While there are many supplementary materials available about applied modeling, usually linked with the use of technology, a couple of respondents mentioned the need for more good commercially available material in this area.

Several of the reform texts have extensive treatments of applied models and of interpreting calculus concepts in applied settings. The "Calculus in Context" text is designed around motivating applications. The Stroyan and Wattenberg texts also contain a large number of applications, and the Harvard Consortium text has a considerable amount. One school built its effort around the "3 Cs" —Cooperate, Communicate, Context. As with projects and writing, the doctoral institutions in the in-depth survey had fewer instances of major new emphasis on modeling and applications. Most reports of increased focus on modeling occurred in conjunction with the use of the Harvard Consortium text.

Impact on Student Performance

As interest in calculus reform has begun to spread beyond those institutions developing reform projects, there has been growing attention to the effect of calculus reform on student learning. People would like data showing that students taught with reform materials do better in some way, e.g., are more likely to continue to the next semester calculus and pass it at higher rates. In the responses to this study's in-depth survey there was a significant undercurrent of concern about the possible harm to students of the decreased attention to algorithmic skills in reform courses.

In the NSF Calculus Initiative, program officers were trying to stretch their funds to support as much development and implementation of new projects as possible, and, when cutbacks were made, one of the casualties was evaluation studies in large projects. Not only did most reform projects have very limited funds to evaluate their impact on students, but a larger problem quickly emerged—there was limited research on assessment and evaluation of student learning in calculus. There was little understanding of how to measure student learning in calculus and more generally in all college mathematics courses. Even simple measures such as retention rate, failure rate, and continuation rate to sequel courses are complicated by confounding factors. For example, retention and pass rates in calculus reform classes might reflect easier tests and/or particularly effective teachers. A workshop was organized by NSF in July 1992 to begin to address these issues. It produced some general guidelines about evaluation in calculus and led to a special working group that was asked to chart in greater detail what future evaluation studies in calculus should try to do.

There are four good studies which provide a comparison of student performance in reform and traditional first-year calculus. Two are well-controlled situations where first-year calculus is split into reform and traditional sections with much or all of the final exam being common to both groups. These studies were undertaken at Baylor and the U. S. Naval Academy in 1993–94, and both involved reform efforts based on the Harvard Consortium text.

Baylor Comparison Study.   At Baylor, the results were controlled for ability by classifying the students in advance into three categories based on high school grades and SAT (or ACT) scores. The final exam had 20 multiple-choice questions based on a traditional course. (There was also a written part of the final that was different for the two versions of the course). The traditional students in all three categories did better on two questions, the reform students in all three categories did better on 11 questions and better in two out of three categories on the other seven questions. One problem favoring traditional students involved related rates that were not covered in the reform text. For students in the weakest category the reform courses produced only marginally better overall scores, but for the average and stronger students, the reform students averaged a full grade higher in the multiple-choice test (e.g., B versus C). The Baylor mathematics faculty had been split about 50–50 on the value of reform at the beginning of the 1993–94 academic year. After seeing this data, the faculty voted unanimously to switch all sections to the reform text in fall 1994. It is worth noting that the students were originally broken into eight ability categories (and later stratified into three categories), and the correlation between ability category and performance in calculus, reform or traditional, was extremely high.

Naval Academy Comparison Study. At the Naval Academy, students were assigned in a random fashion to reform and traditional sections. The final had 10 common questions, four graphical in nature and six traditional. Traditionally-taught students did a bit better on one question; the reform students did a bit or a lot better on the other nine. One traditionally-based problem on which the reform students far outperformed the traditionally-taught students was the following: find $h'(2)$, where $h(x) = g(f(x))$, $g(x) = x^2$ and values are given for $f(2)$ and $f'(2)$.

Project CALC Evaluation Study. Project CALC at Duke is the one large NSF reform project with a substantial evaluation component, run by Jack Bookman. This project involves heavy use of computers, exploratory group learning, and writing. There were many student complaints in the first two years that were largely addressed in subsequent years. Since fall 1992, Project CALC has been used in all first-year calculus classes at Duke. In the second and third years of the project, spring 1991 and spring 1992, Bookman gave a problem-solving test to selected samples of Project CALC and traditional calculus students (who used Thomas and Finney) at Duke. The tests emphasized problems in both applied and purely mathematical settings. The Project CALC students' performances were better both years by statistically significant amounts (a 20% higher score for Project CALC than the traditional course).

In spring 1992, he gave tests to selected groups of sophomores and juniors who had had Project CALC and traditional calculus. One part of the test probed experiences based on calculus, such as "I've applied what I've learned in calculus in other courses"; and "after I've forgotten all the formulas, I will still be able to use the ideas of calculus". Students were asked to answer on a scale from 4 to 1 measuring their level of agreement with the statement (higher was better). Project CALC students had 20% higher scores, a statistically significant difference. On a skills test, e.g., integration by parts, the traditional students did better but not significantly better. On a problem-solving test, Project CALC students did better but not significantly better.

Since Project CALC was instituted for all sections, the continuation from Calculus I to Calculus II has increased about 10%. Data from early Project CALC students shows that they took on average one more

mathematically-based science/engineering course than traditionally-taught students took and were more likely to take additional mathematics courses. The GPA in subsequent mathematics and sciences courses was the same for reform and non-reform Duke students.

Purdue C^4L Study. At Purdue, some sophisticated statistical analyses of subsequent enrollment in calculus and post-calculus courses have been undertaken. The study compares students whose first calculus course was reformed (the Purdue C^4L project course) with those whose first course was non-reformed. Allowing for choice of major, predicted GPA, and other factors, the study found that the average number of subsequent semesters of calculus and the average number of post-calculus courses was greater by a statistically significant amount for students in the reformed course.

General Indicators of Student Performance. Through this MAA assessment study's own surveys and by drawing on articles and studies by others, the following somewhat more broadly based qualitative assessment of student performance has been assembled.

First, we note that the dearth of literature on evaluation of learning in undergraduate mathematics courses reflects traditions in higher education that give considerable responsibility to individual departments and instructors for the structure of their own courses and that recognize that the value of a college course, and more generally a college education, are very difficult to quantify. In light of these factors, it might be argued that a systematic evaluation of reformed calculus or a comparison of student learning in traditional and reform courses tries to create and employ measures of student learning that faculty do not really want. However, the desire to know in advance whether the greater effort required in teaching a reformed calculus course does more good than harm has generated much interest in evaluation of calculus reform.

This study's in-depth survey asked faculty for their assessments of changes in student performance and enrollment arising from calculus reform. Many responses noted that reform efforts had just started and it was too soon to discern any changes. However, a quarter of the respondents indicated that reform students were doing better in terms of passing rates and general performance. A quarter also felt that reform courses improved retention rates in calculus. A fifth of the respondents said that post-calculus enrollments seemed to have increased after calculus reform was instituted, although for most schools reform was too young to have many sophomores or juniors who had taken reform calculus. Some noted that group work in calculus courses seems to be promoting more out-of-class group studying in mathematics and other subjects.

Two-year college respondents noted that there had been significant gains in retention, although their numbers were small. One campus attributed this to the strength of the cooperative learning approach. The bachelor-degree level institutions noted a shift away from failures and A's. The distribution seems to becoming a bit more bi-modal, but at the B/C levels. At the master- and doctoral-level schools less was reported on the effect of on student achievement. One respondent reported that it tends to "smooth" student performance. The university responses tend to attribute better retention and pass rates to the fact that reform calculus appears less threatening. A few respondents from institutions with reform in place for a couple of years mentioned that the number of mathematics majors had increased. Unfortunately, no respondent cited careful studies about pass rates or retention or increased advanced enrollments.

When asked what their students were learning now that they had not learned before, the respondents did not have solid evidence to indicate that their students were significantly different in terms of the amount of calculus content they had learned and retained. However, they did feel, on the basis of case evidence and anecdotes, that their reform students were stronger in calculus. Again, there was a dearth of informative responses about changes in knowledge or skills from the doctoral institutions.

Reasoning Skills and Other Abilities. Half of the 62 in-depth survey respondents indicated that they perceived improved conceptual understanding in students, and a quarter of the respondents said that students' general problem-solving skills had increased. The 1993 in-depth survey answered by 18 institutions asked for specific information about changes in students' process abilities. Many respondents said that they did notice some such effects of the reform efforts. Several respondents at the associate-, bachelor-, and master-degree levels indicated that their students were becoming more proficient in their attacks on and solutions of open-ended problems. Overall they reported a noticeable increase in student skills to communicate and reason mathematically, and to attack contextual problems. Doctoral level respondents in this survey reported observing little if any changes in process abilities.

Differential Effects of Reform. There was little agreement on what type of student benefits more from reform approaches. Many faculty noted that the stronger students seemed to get more out of reform courses. Others felt that strong and weak students benefited equally from reform approaches. Several faculty have said that the conceptual emphasis in reform courses is more forgiving of weaknesses in algebraic skills, especially for older returning students, and allows part-time and older students with such deficiencies a chance to succeed in calculus. Several noted that B-level students often got deeply engaged by reform methods and earned solid A's. There have been reports, one well documented at SUNY-Stony Brook, that reform courses have worked very well with Treisman-type workshop programs to assist underrepresented minorities in excelling (earning A's and B's) in calculus.

Comment from Client Disciplines. The survey asked about feedback from client departments, but most respondents indicated no feedback to date. Seven did note that client faculty had expressed substantial pleasure with the technology facility students had gained in reform calculus courses, and five said client faculty had commented that reform students appear to have better reasoning skills. Four cited negative feedback that reform students seemed to have weaker algebraic skills. Information from reform text developers indicate that reform texts, particularly the Harvard Consortium text, have been faring better with engineering departments than anticipated. Engineering accreditation committees have been overwhelmingly supportive of reform approaches.

Testing. Aspects of course administration, such as testing, have changed little to reflect the different type of learning expected of students in reform courses. Several respondents to this study's in-depth survey indicated the need to lessen the heavy reliance on timed tests, but find themselves bound by the overall class schedules or commuting patterns of their students. One school reported shifting 35 percent of the evaluation to out-of-class activities. Others indicated the development of "gateway" type exams as adjuncts to the course to deal with the testing of mastery and prerequisite knowledge in order that classes could focus on the concepts and procedural aspects in a contextual setting. Much remains to be done about evaluation and testing of student learning in reform calculus classes.

Impact on Student Attitudes

The in-depth survey asked respondents to characterize their students' reactions to the changes that had been made. Overall these reports were positive. One quarter of the 62 respondents stated that students showed noticeably greater self-confidence in their mathematical reasoning ability and skill in attacking open-ended problems. One-quarter cited the pleasure students showed in becoming proficient in the use of technology. A good number wrote that students seemed to find mathematics more interesting and more useful. Most commented that students reported spending more time on the course than in the past. While there were complaints about the extra work, there did not appear to be any major revolts. A surprise was that there were few reports of student complaints about too many "word problems." The 1993 survey asked respondents to speculate on the source of the changes noted in students in reform courses. The responses generally highlighted one of two sources – technology or the changes in actual classroom teaching methods.

A number of respondents spoke about how students did not notice when laboratory or recitation periods ended in reform courses. Bookman's observational studies at Duke documented that, while falling asleep and other signs of inattention were common in traditional Duke calculus classes, they were extremely rare in Project CALC classes.

Impact of Technology. Students report that technology use was more time consuming, but they liked the portability of the graphing calculators. The role of technology in calculus reform seems to have had unexpected benefits in student attitudes towards mathematics. While calculus and the high school mathematics that precede it deal with ideas hundreds or thousands of years old, in comparison to the 20th century ideas discussed in science courses, mathematics has become a leader for most students in the use of technology. The proficiency with graphing calculators and CAS software they acquire in reform calculus classes can be used to advantage in other quantitatively oriented courses. The power and diverse uses of technology become associated with mathematics in students' minds. Thus mathematics is seen as contemporary and as a tool for the future. The use of calculators and computers allows for alternative methods of viewing a problem, of bringing motivating real world problems into the classroom, and of enabling mathematics to "make sense" for students.

Most Students Overcome Challenges of Reform. Respondents indicated that programs instituting change have to be prepared for implementation-lag on the part of students. Most reported that students had difficulty in adjusting to cooperative learning, problem sets, writing, and projects. On the other hand, students adapted fairly quickly to technology use, especially where they saw that it was playing an integral role in the course. They did not like paying for non-used calculators! Several saw the efforts to bring context and rationale to the study of calculus as providing students with a course which was "not detached from their goals and interests."

According to students, aspects that make reform courses harder are:

- the necessity to read the book;

- problem sessions outside of class; and

- the absence of a visible teacher/leader spooning out information.

Reports fairly consistently indicate that resistance to cooperative learning, writing, and project work has been due in large measure to students' inexperience with learning mathematics in these formats. On the whole, the reports indicated that by the end of the first semester, and surely by the end of the second semester, most students had adapted to the expectations of the course and that overall there was widespread satisfaction with these elements. Respondents generally commented that, even through this period of shock and resistance, students were stronger in problem-solving and conceptual understanding from the outset.

Some faculty have expressed concerns about potential difficulties for students transferring between institutions with different types of calculus courses, but there is no evidence that this has been a problem. It is surprising that there was virtually no mention of problems with students moving in successive semesters between reformed and traditional courses, except for a couple of references to reform-taught first-year students wanting reform continued into their sophomore calculus courses.

Students Resistant to Reform. The changes in pedagogy and the new emphasis on conceptual understanding have produced resistance and negative reaction from students who don't want to have their "comfortable relationship" with mathematics disturbed – even if that relationship is distasteful. In other words, some students prefer to have a course devoted more to building rote skill rather than to gaining a deeper understanding of "what is going on" and fostering active learning through activities with technology, writing, and cooperative groups. Hopefully, as the influence of the NCTM Standards on school curriculum and instruction grows, fewer entering college students will display this attitude.

Probing whether there were differential effects in students' difficulty with reform elicited these comments:

- AP students and repeaters in calculus were more resistant and more challenged than students taking the course for the first time.

- Most of the students who fought reform to the end were students who relied on memorization or rejected the use of technology for visualization.

- Some students who had struggled in mathematics bloomed in an exploratory or writing-oriented program.

Students at more prestigious institutions who did not take calculus in high school have been some of the most vocal and articulate critics of reform. These students tend to be pre-professional, e.g., pre-med, and getting them to stop and think in calculus is a challenge that probably does require radical reform. *Math Horizons,* the new MAA publication for undergraduate students, had an article in its premier issue, "Does Calculus Need Reform?," with interviews from several students "speaking out" on this question. The comments of Daniel Grossman, a junior at Harvard, about the Harvard Consortium reform effort reflected the concerns of several interviewees. He said, "the reform is fundamentally flawed in that it is not really reform at all. The computer component and the graphing calculator and the supplementary problems all depend on the enthusiasm of the students to be effective; but students perceive anything outside the primary tools of the course to be an extra burden, and that kills their enthusiasm. The only way to make these tools effective is to make them the primary component of the course—that would constitute radical reform that most departments are not really willing to undertake (*Math Horizons,* first issue, page 12)." The eventual student acceptance of the more radical Project CALC approach at Duke supports this point of view.

Impact on Faculty Attitudes towards Change

The large attendance at talks, mini-courses, and contributed paper sessions at professional meetings and at calculus conferences attests to a strong curiosity among thousands of mathematicians about calculus reform.

How Reform Occurs. The introduction of calculus reform at most institutions occurs as follows. A small number of faculty become very much interested in reform and attend professional meetings or special conferences where they learn about reform efforts and talk to people at other institutions interested in reform. They come back to their departments highly motivated to institute change and, after a period of extensive preparation, teach some experimental reform sections. Some of them get funding for Calculus Initiative or ILI projects or obtain local funding for equipment. The in-depth survey data indicated that in middle-sized departments implementing some level of reform, faculty had collectively attended about 10 presentations on calculus reform (at professional meetings, conferences, short courses, etc.) when beginning to experiment with reform. Based on results of experimental sections, a reading of reform materials, and information from *UME Trends,* MAA periodicals and word-of-mouth about reform activities elsewhere, other faculty become interested in reform and increase the number of experimental sections or endorse a decision for course-wide adoption of reform.

At all types of institutions, there was usually considerable skepticism and open opposition to reform initially. In time most faculty accept, or at least allow others to implement, calculus reform. However, there is often a minority of faculty who remain uninterested in, or opposed to, reform.

Sixteen of the institutions, a majority of those implementing major reforms in the in-depth survey, reported that they had organized workshops and short courses to prepare faculty in reform materials and pedagogy before course-wide reform was implemented. At the nondoctoral institutions, the faculty development programs seemed to be centered mostly around faculty seminar groups, with some input from outside sources. Several institutions of all types reported weekly seminars to plan and support the reform efforts.

Faculty attitudes towards reform are necessarily influenced heavily by what level of reform is being proposed at an institution. For example, while some faculty might resist reform at the level of the Harvard Consortium text, other faculty might argue that the Consortium text is so modest in its goals that it does not constitute reform. Not surprisingly, the vast majority of reform courses are at the more modest level.

The in-depth survey asked a number of questions to try to learn about faculty attitudes towards calculus reform and how change occurred in individual departments. Table 9 gives data about which faculty, junior or senior, were more active in supporting change. Early reports had

Table 9: 1993, 1994 In-depth Surveys of Faculty Interest in Reform at 62 Institutions

Type of Institution	Senior Faculty leading change	Junior Faculty leading change	Most Faculty Support Change
Doctoral	1	2	5
Master's	1	1	9
Bachelor's	1	4	17
Two-Year	2	0	8

[11 Institutions had no faculty or a small number of faculty supporting change.]

indicated that most change was coming from senior faculty, but Table 9 provides no support for that view.

Table 10 lists commonly cited hurdles to reform mentioned in the in-depth survey.

Table 10: 1993, 1994 In-depth Surveys of Hurdles to Calculus Reform at 62 Institutions

Difficulty in getting computer equipment	18
Resistance of some faculty to changing established ways of teaching calculus	12
Little faculty interest in instructional issues	3
Faculty felt reform was too time-consuming	3

Master's degree institutions with graduate assistants indicated that the involvement of graduate assistants tended to be fairly low, although these institutions may use TAs for precalculus courses. Doctoral institutions reported varying levels of TA involvement with reform. There have been several cases reported where TAs have tried to avoid reform sections because they took more time, time away from dissertation research.

When reform is adopted course-wide at two-year, four-year, and master's degree institutions, it appears to have gained quite broad faculty support, while at doctoral institutions there often appears to be continuing reluctance among a significant fraction of the faculty to adopt reform approaches. This pattern may be explained by the fact that their faculty are typically more closely involved in calculus teaching and undergraduate education issues generally than at doctoral institutions. At two-year, four-year, and Master's degree institutions the in-depth surveys found that there tends to be a consensus-building process before reform is adopted course-wide.

At institutions that have not adopted reform course-wide, there are typically a fair number of faculty who

are wary of the claims made on behalf of calculus reform. However, one great advantage reform enjoys is that there are few faculty who are strong defenders of the current situation in calculus. The emphasis on pedagogical changes has also made it hard to attack reform directly. Most faculty agree that projects, increased writing, cooperative learning, and closer connections to client disciplines are desirable goals.

All respondents indicated that reform teaching takes more time in grading project papers, meeting with students, dealing with technology, and planning for class. Developing a reform course takes a tremendous amount of time. It was rare for faculty members involved in developing or implementing calculus reform, unless supported by NSF funding or other sources, to have obtained release time for their efforts.

Impact on Faculty Teaching Reform Courses. A common theme in responses to the reform's impact on faculty was that those faculty actively engaged in teaching reform courses feel very good about what they are doing. Although limited changes had been seen to date in many cases, the faculty still reported that they were sure that what they were doing was right for students and right for themselves as mathematics professors. There were no responses saying, "I hope this effort turns out well because I have put a lot of effort into it." Rather, there was a deeply held sense that no matter what the short term results might be the general goals and approaches were right. The concerns of faculty engaged in reform focused on issues such as what is the appropriate balance of concepts and skills in first-year calculus.

In traditional calculus courses in MS and doctoral institutions, calculus has often been taught with common final exams following a highly structured weekly course outline. This framework provided limited opportunity for the lecturer to be actively engaged in the course instruction. On the other hand, reform courses tend to give the instructor, as well as the student, a more active role in what happens in the classroom (and outside it). This is a framework in which teaching can be much more satisfying, even if there is a common final exam.

Factors Affecting the Reform Process. Many respondents reported that the encouragement and active support of the department chair and the campus administration were important factors in successful implementation of reform efforts. This support took important tangible forms such as travel money for calculus conferences and equipment funds. While there are stories about some departments where a couple of reformers were opposed by the chair and most other faculty, this survey's responses showed that chairs were quite consistently ahead of the average faculty member in their receptiveness to reform. The finding is consistent with the 1994 report of the JPBM Committee on Professional Recognition and Rewards (JPBM, Washington, 1994) which reported that department chairs tended to be more sensitive to the importance of good teaching and curriculum reform than their faculties.

It is not clear how important the involvement of the universities in the Harvard Consortium—Harvard, Stanford and Arizona—has been in helping calculus reform gain faculty acceptance in doctoral institutions. While the Consortium text is the most popular reform text at doctoral institutions(as it is at the other levels as well), research universities often use "home-made" reform—supplementary materials in problem sections or labs—as often, and over a dozen are using the Ostebee/Zorn text (developed at a four-year college).

When asked about the amount of outside contact they made with client disciplines during the planning and implementing of their reform sequences, several doctoral institutions with reform programs indicated that they had significant consultation outside their departments. At smaller institutions, the amount of client input was uniformly distributed among "some," "little," and "none". Departments most often noted as having been consulted were in engineering, computer science, physics, chemistry, economics, and business.

Future Acceptance of Reform. At the writing of this report in mid-1994, one finds a growing sentiment at two-year, four-year and master's institutions that calculus reform is inevitable and that it is just a matter of deciding when to do something and what level of reform to choose. Because faculty know that reform will require increased effort, they are often in no hurry to undertake reform. At doctoral institutions, there is a large middle

ground of undecided faculty who have not given much thought to calculus reform and are not interested in becoming more deeply engaged in calculus instruction.

When proponents of calculus reform look for potential future problems in its acceptance within the mathematics community, the word that always arises is *backlash*. Few can conceive of a split world with one set of institutions offering reform courses and another set offering traditional courses. Rather the fear is that the wave of reform will be reversed at many institutions and reform courses dropped. As the pace of reform accelerates, course-wide adoption undertaken without adequate groundwork among faculty may lead to unhappy faculty and students who rebel and insist that reform be undone. In the first year of implementation of the Harvard Consortium text, potential test site instructors were questioned and background investigations were conducted worthy of a Top Secret Clearance. The reason was fear that the materials would be poorly used, thus creating a bad image for the project. With hundreds of satisfied institutions, many highly regarded, currently using the Consortium Calculus text, if institution X now has a bad experience with it, the problem will no longer be ascribed to the Consortium text. However, this fear still haunts reform developers.

There have been a small number of institutions where backlash has occurred with reform texts and other efforts. However, the drop rate for reform texts is well under 5%. Realistically, if backlash was going to be a serious problem, it is much more likely that it would have occurred early in the reform movement when materials were in rough shape and reform was viewed as a risky undertaking.

Extended Impact at Colleges and Universities

Impact on Other Mathematics Courses. The reform movement in calculus appears to be spreading to other mathematics courses. Many respondents to the in-depth survey indicated that they were looking to implement a graphical/functional viewpoint in precalculus and that technology was being introduced into other courses, most commonly precalculus, linear algebra, and differential equations. At the bachelor's and master's institutions, there were reports of the use of computer algebra systems (CAS) in upper division courses as a result of their usefulness in calculus. After using CAS in calculus, one respondent said, "I cannot wait to teach complex analysis next year using CAS." There were also reports of greater use of discussion rather than lecture in the classes. A few doctoral-level institutions reported that change occurred first in differential equations and linear algebra and then moved down to calculus.

Impact on Faculty Morale. One of the most heartening findings from the survey was the feeling of growth and professionalism among faculty at non-doctoral institutions. Many of these respondents reported that calculus reform raised morale among the mathematics faculty. Some respondents made statements to the effect that participation in the calculus reform has revitalized the department. Others noted that the faculty were finally moving to become "computer literate" because of the reform efforts on their campuses.

Growing Interest in Pedagogy. The calculus reform movement has increased interest in pedagogical issues among mathematics faculty. More faculty are lecturing less and exploring other modes of instruction. They are experimenting with other modes of assessing student learning besides the standard pencil-paper timed tests. While these efforts are time-consuming, many faculty are excited about trying to improve instruction and evaluation.

This interest in pedagogy has raised interest in the larger goals of mathematics courses, looking at the forest, not the trees. Reform calculus courses are trying to teach aspects of mathematics that are not specific to calculus.

Research in College Mathematics Education. Calculus reform has also helped raise interest in educational research into how undergraduate students learn mathematics. The MAA recently began having a major plenary talk on research in undergraduate mathematics education at the national AMS/MAA Joint Meetings, and audiences have been generally larger than those for mathematical research talks. A joint AMS/MAA Committee on Research in Undergraduate Mathematics Education was recently formed and is publishing annual volumes in this area. Such professional society interest in education research was unimaginable a few years before.

The promising start of the calculus reform movement has led to calls for research and educational yardsticks to evaluate learning in calculus, both in traditional and reform courses, in terms of the larger goals of a mathematical education. A special NSF workshop was convened in July 1992 whose purpose might informally be summarized as, "if calculus reform is succeeding, what is it that students should be learning and how will we know?" The workshop agreed on the following objectives for good calculus instruction:

1. transfer techniques of calculus to other disciplines and novel situations;

2. reason analytically and communicate mathematical ideas symbolically, in writing and orally;

3. appropriately use technology in solving problems;

4. deal with complex, often ill-defined problems;

5. become independent mathematics learners;

6. translate problems from one form to another;

7. work effectively in groups;

8. represent problems in more than one way: algorithmically, graphically, numerically, symbolically, and verbally;

9. appreciate the role of experimentation, conjecture, verification, and abstraction in mathematics;

10. develop positive attitudes about one's ability to do mathematics successfully, appreciate its elegance and structure, and pursue further work in quantitative fields;

11. reason qualitatively and quantitatively; and

12. develop in-depth understanding of specific mathematical concepts

Assessing how well a calculus course meets these objectives is a daunting task. The workshop led to the formation of an NSF working group on assessment in calculus. The report of that working group, chaired by Alan Schoenfeld, will be available in the near future.

Impact on Attitudes Towards Mathematics Departments Across Campuses.

> Science cannot live by science alone. Research needs education, just as education thrives when it is conducted in an atmosphere of inquiry and discovery. In fact, the separation of education and research makes no sense intellectually. It is an artificial and unhelpful separation caused by how many of us in higher education have chosen to behave.

> —Neil Lane, Director, National Science Foundation, *Science in the National Interest*

Calculus reform has changed not only how many mathematicians look at their undergraduate teaching, it has also changed how others in academia look at mathematics departments. At doctoral institutions, mathematics senior faculty were perceived to have little interest in calculus instruction. Mathematics departments used to have a reputation of being one of the most tradition-bound and low-technology units on campus. Foreign languages and political science were making more use of computers than mathematics departments. That has changed significantly on campuses involved in calculus reform.

Administrators and other departments know how difficult it is to make a major change in a large introductory course. Counting all tracks and all semesters, calculus is the biggest of introductory college subjects, and the most important for other disciplines. Most reforms in large introductory college courses are pet projects of a few faculty and are abandoned in a year or two. Thus the rest of the campus is impressed by, and respectful of, mathematics departments both for undertaking a major change and for obtaining generally positive results. Although many mathematicians are uncertain about long-term prospects for calculus reform, its current image is generally one of success, and mathematics departments are benefiting from this image.

In these tight fiscal times, it is hard to measure how much impact calculus reform has had on helping mathematics departments retain or add faculty. Discussions with deans indicate that mathematics departments involved in reform have gotten some favorable treatment. The in-depth survey indicated that local institutional support for computing equipment for mathematics has increased with calculus reform. There are some institutions that have given new resources to mathematics departments to hire additional faculty as part of calculus reform implementation efforts. Most notable has been the University of Michigan, where upwards of $1,000,000 was given for new mathematics faculty as part of an initiative to put more faculty into calculus classes in order to improve student learning and reduce failure and dropout rates.

Increased Attention to Instruction in Tenure Decisions. Another result of calculus reform, and the associated increased attention to instruction, is that mathematics faculty have earned tenure at some leading research institutions, including Harvard and University of Michigan, for accomplishments in the educational domain. Indeed, the Michigan Mathematics Department's tenuring of faculty for educational excellence was part of the commitment to good teaching and calculus reform that helped convince the university administration to invest major resources in mathematics.

Impact on School Mathematics Teaching and NCTM Standards. Calculus reform has been widely applauded by faculty interested in the preparation of teachers of school mathematics. There is much evidence that K–12 teachers tend to teach the way their own teachers taught them. Despite wide support of the NCTM *Standards* in academia, until recent implementation efforts in calculus reform began, college and university faculty had probably made few changes in their own mathematics instruction in support of the NCTM *Standards.*

Although independently developed, the NCTM *Curriculum and Evaluation Standards for School Mathematics* and the *Professional Standards for Teaching Mathematics* have much in common with calculus reform efforts. Both emphasize conceptual understanding and making students become active learners through exploratory learning, projects, and technology. Curriculum and teaching along the lines of the NCTM *Standards* are excellent preparation for reform calculus courses. Conversely, for secondary schools to prepare their students well for calculus taught in a reform style, secondary schools need to adopt curriculum and teaching styles similar to those advocated by the NCTM documents.

One of the greatest hurdles that calculus reform has faced has been habits of rote memorization acquired by students in high school courses that were seeking to prepare students for traditional calculus courses. Current high school practices, especially in precalculus, will be slow to change towards the NCTM *Standards* until colleges demonstrate that they want students prepared for reform courses rather than for the more traditional programs with placement tests based in heavy symbolic processing tasks. In addition to the better pedagogy in reform calculus classes, the reform movement is stimulating efforts to make students more active learners in upper-division mathematics courses taken by prospective teachers.

Impact of Use of Technology. Unlike the use of computers in language labs for practice drill, the use of graphing calculators and computers introduced in calculus classes develops skills with technology that are applicable in other quantitative courses. For this reason, mathematics is now acquiring a reputation among students and faculty as being a leader in the use of technology on campus. This image has facilitated efforts to acquire computing equipment and support staff for mathematics departments. ILI grants to mathematics departments have increased since the NSF Calculus Initiative in a symbiotic fashion: reform motivates and justifies the need for technology while technology forces a rethinking of what topics are taught and how they are taught. Still, use of technology in postsecondary institutions lags behind its use in high school mathematics programs.

New NSF Collegiate Mathematics Education Initiative. The new NSF initiative, "Mathematical Sciences and Their Applications Throughout the Curriculum," is one of two major new comprehensive initiatives in collegiate education at NSF. (The other is in chemistry.) This initiative is a recognition of the success of the calculus reform movement. Its goal is to share the spirit of reform developed in the mathematics community during the Calculus Initiative with other disciplines. Developing projects is expected to involve close cooperation among faculty in the mathematical sciences and other disciplines. Concepts and topics from the mathematical sciences are to be incorporated in these other disciplines, and the perspectives of the other disciplines are to be used in the mathematical sciences courses as well. This interdisciplinary perspective is central to the new initiative and presents a considerable challenge to the community. Proposals are expected to show involvement by a multidisciplinary project team and to have institutional support across several academic units. In the June 1994 first call for planning grants, the program received 191 proposals. Clearly the community is actively thinking about curricular reform and is anxious to move ahead with the spirit of reform to other aspects of the undergraduate curriculum.

New GRE Mathematics Aptitude Test. The most significant impact of the calculus reform to date outside of the mathematics community is the recent decision of the Graduate Record Examination program to develop a new mathematics reasoning general (aptitude) test which will contain questions in the spirit of calculus reform. The general aptitude part of the Graduate Record Examination will be reorganized in fall 1997 to consist of five parts: (i) the current quantitative test, (ii) a revised verbal test, (iii) a revised analytical reasoning test, (iv) a new mathematical reasoning test, and (v) a new writing sample test.

The current GRE quantitative test assumes only two years of high school algebra, to accommodate humanities students who have had no mathematics since high school, and produces uniformly high scores among college seniors in mathematically oriented disciplines. These high scores limit the value of the test for admission decisions in mathematically oriented graduate programs. (At the same time, non-mathematical graduate programs are also unhappy with it because many institutions base university-wide fellowships on GRE aptitude tests whose high quantitative scores are felt to favor mathematically oriented disciplines.)

The new GRE mathematical reasoning test is expected to replace the current quantitative test for students planning graduate study in engineering, the physical sciences, the mathematical sciences, computer science, economics, and some areas of biology. It will assume that examinees have had a year of college calculus. The new test is still an aptitude test, and so its calculus prerequisite will not result in questions about calculus techniques. Rather, the new test will contain a number of questions that involve the ideas from precalculus and calculus, commonly in applied settings. The staff of the Educational Testing Service (ETS) proposed to the committee of experts from quantitative disciplines assembled to specify the content of the new test that questions in the spirit of calculus reform be used. When shown a sample of such questions, the committee was very enthusiastic and voiced the belief that this new test would come to carry substantial weight in admissions decisions in their disciplines. Below are three sample questions prepared by ETS staff.

1. The graph shows a manufacturer's profit P as a function of x, the number of items produced and sold. At which of the eight marked points on the graph does the profit per item have the greatest

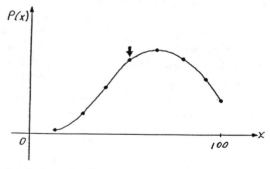

value? Answer: D

2. The graph shows the concentration of a drug in the bloodstream of a patient during a medical treatment over a 12-hour period. Of the 12 one-hour intervals on the time axis, highlight the one interval during which the concentration increased most rapidly.

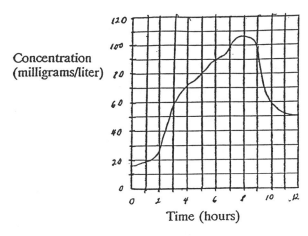

Time (hours)

Answer: (2, 3)

3. The line tangent to the curve $y = Ce^{-kt}$ at $(0, C)$ intersects the t-axis at $t = T$ as shown in the figure.

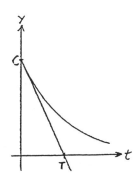

What is k in terms of T? Answer: $k = 1/T$

Thus, the new mathematics aptitude test for quantitative disciplines will be heavily based on calculus reform. The combination of this test and the new graphing calculator-based AP calculus test will create substantial pressure for unreformed institutions to switch to calculus reform. Students who are not proficient in calculus reform ideas may have to undertake extensive preparation, possibly even a Stanley Kaplan reform calculus review course.

Impact of the NSF Calculus Initiative

From the start of serious efforts to develop reform calculus courses, there has been universal agreement that the calculus reform movement, as we now know it, would have been impossible without the NSF Calculus Initiative.

Of course, the readiness of mathematics faculty to consider reform was also a critical factor, and there were other important developments that anticipated and supported calculus reform, mentioned in section II.2. The

1993 in-depth survey included a question about the role of the NSF Calculus Initiative in the reform movement. It seems fitting to close this report with some typical responses about this centerpiece of the calculus reform movement.

There is absolutely no doubt that the NSF Calculus Initiative was fundamental to the changes that have taken place in our program. Quite frankly we had no idea how complex the task of changing our culture would be and no idea of how much time and energy it would consume. The NSF program supported a number of models and materials development programs. From some of these (e.g., St. Olaf) we have drawn extensively, from others (unnamed) we have seen what to avoid. However all of them have been useful. Finally, the financial support the program provided has been pivotal to actually getting the changes made. Without it we simply would not have had the resources to carefully adapt the extant models to one that best suits our situation. Absolutely, because without it I doubt whether the movement would have been able to build the momentum it has now produced. I am a bit worried about the use of the past tense in reference to the NSF calculus initiative, however. The job of reform is FAR from finished, and it is to be hoped that the program continues! Yes, major funding has resulted in major change here, reaching far beyond calculus. Now the changes in calculus are affecting much of our undergraduate and graduate teaching.

It was essential at our institution, where research is heavily emphasized for promotion, tenure, and salary. Receiving two NSF grants made it possible for me to devote the necessary time (i.e., all of it!) to developing our laboratory program without giving up all hope of promotion.

The NSF initiative was a very positive force at our college. It got us involved in the ACM/GLCA project. NSF, though not the Calculus Initiative, also funded our PC lab through the ILI program—that really got us going! The NSF initiative also got me to apply for a Calculus Implementation grant. Even though we did not get this grant (and I was not pleased at the rejection), the application effort forced us to think through how we were going to proceed and we eventually followed this plan pretty well."

Yes, yes, yes. There has been awareness of the need for a change for about 20 years, but the materials weren't there. I'm sure they will continue to evolve, but the quantum leap has been made, thanks to NSF's support.

Closing Remarks

This study assessing the calculus reform effort did not ask in its surveys what remains to be done. Likewise, it did not ask respondents whether they thought calculus reform is a success or will be universally implemented in some form by the year 2000. However, there is growing sentiment in many departments not now fully converted to reform, that reform is becoming inevitable. The pace of reform is so fast that spring 1994 predictions will probably be dated by spring 1995.

One can safely say that there were very few individuals at the time of the Tulane Conference who would have predicted the amount of change generated thus far by the calculus reform movement. Even those who attended the conference would have considered the present extent of reform beyond all reasonable expectation. One of their goals, stated in the Tulane Conference Implementation Workshop report, was to make reformed calculus "the property of the entire mathematics community and not just a program of the New Orleans conference or of those involved in the initial projects." Through the efforts of the NSF Calculus Initiative, particularly its emphasis on dissemination and implementation, calculus reform is truly a property of the entire mathematics community.

Appendix A
Background Information about In-depth Survey

Table 1: Data about Institutions in In-depth Surveys

Institution/Size	Average Total Enrollment	Average Number of Faculty	Average Calculus Enrollment
Doctoral	20,000	52	4100
Master's	9,000	25	1160
Bachelor's	3400	12	360
Two-Year	4500	10	510

Participating Institutions in the 1993 in-depth survey

Doctoral Degree Institutions (4) :
> University of California-Davis, Clemson University, University of Maryland-College Park, Michigan State University.

Master's Degree Institutions (5):
> American University, University of Maine-Orono, Portland State (OR), University of South Dakota, Western Washington University.

Bachelor's Degree Institutions (6):
> Canisius College, Greenville College (IL), Grinnell College, James Madison University, Loyola College (MD), University of Wisconsin-Oshkosh.

Associate's Degree Institutions (3):
> Bethany Lutheran College, Hutchinson Community College, Meridan Community College.

Participating Institutions in the 1994 in-depth survey

Doctoral Degree Institutions (7):
> University of Arizona, Baylor University, Brigham Young University, University of Hawaii-Manoa, University of Missouri-Rolla, North Dakota State University, University of Washington.

Master's Degree Institutions (8):

California State University-Bakersfield, Central Missouri State University, John Carroll University, Miami University of Ohio, Oakland University, University of Puerto Rico-Mayaguez, Southern University, University of Southern Mississippi.

Bachelor's Degree Institutions (20):

Armstrong State College, Davidson College, Doane College (NE), Fort Lewis College, State University of New York-College at Geneseo, Houston University-Downtown Campus, Knox College, Lafayette College, Macalester College, Mary Washington College, Massachusetts Maritime College, Meredith College, Northern Kentucky University, Oral Roberts University, St. Mary's College (MD), Santa Clara University, Seattle University, Southern Utah University, Waterville College, West Virginia Wesleyan College.

Associate's Degree Institutions (9):

Austin Community College, Bismarck State College, Broome Community College, Delaware Technical College, Jones County Community College, Mission Community College, Normandale Community College, Pearls River Community College, Pueblo Community College.

Appendix B

1994 Short Survey
and
1994 In-Depth Survey

Survey

MATHEMATICAL SCIENCES DEPARTMENTS OR PROGRAMS

Name: _____

Institution: _____

Phone: _____   E-mail:_____

Category of Institution (Please check exactly one category):

_____ Two-year College   _____ Four-year College

_____ Master's degree granting   _____ PhD granting

1.   To what extent has the calculus reform movement led to a change in the calculus sequence at your institution over the past three years?

___ no appreciable change   ___ modest change   ___ major revisions

2.   If there have been changes, do those changes involve (check whatever applies)

___ experimental sections   ___ all calculus offerings

3.   Approximately how many faculty, graduate teaching assistants, and students are participating in revised calculus offerings each <u>year</u>?

Group	Number	Percent
Faculty		
GTA's		
Students		

4.   Identify any sources of funding your department may have received (check all that apply):

_____ National Science Foundation
_____ Other Federal Agency (e.g., FIPSE)
_____ State
_____ College or university
_____ Other (describe)

Please return by March 1 to:
Mathematical Association of America, Attn: Jane Heckler
1529 18th St., NW, Washington, DC 20036

ASSESSING CALCULUS REFORM EFFORTS (ACRE)

Institutional Survey

1. **Basic Background**

   a) What is the approximate undergraduate enrollment at your institution?

   b) How many (full-time) faculty in the mathematics department? How many part-time faculty?

   c) What is the approximate annual enrollment in all your calculus courses?

   d) Which first-year calculus courses (e.g., calculus for math/science majors, business calculus) are using a 'reform calculus' approach and what are the approximate enrollments (if not all sections of a course are using reform approach, estimate percentage using reform calculus)?

   e) How do you expect the number of students taking reform calculus to change over the next two years?

2. **Changes in Calculus**

   a) To what degree has each of the following features been a component of reformed calculus courses:

   - i)   Calculators or computers
   - ii)   Cooperative learning groups
   - iii) Student projects
   - iv)   Writing/communication emphases
   - v)   Applications/modeling.

   b) How has the emphasis placed on concepts, on techniques, and on solving more complex problems (often applications related) changed in reform courses? What changes have there been in the selection of topics covered? What have been the major decreases in emphasis? major increases in emphasis?

   c) What impact has the reform in calculus had on other coursework offered in the department? (such as items mentioned in a), b) above)

3. **Student Achievement and Attitude**

   a) Has there been a change in faculty perceptions of students' understanding of calculus concepts and problem solving/reasoning skills in reform calculus courses?

   b) How have success (C or better) and retention rates changed in reform calculus courses? Are more students taking advanced mathematics courses or majoring in mathematics?

c)   What changes are perceived by faculty in students' attitudes toward the calculus (e.g., harder (longer) to do, more interesting) or toward their own self-concept as a student of mathematics? This applies to both how reform calculus is taught and what is taught.

d)   What comments do you have from faculty in client disciplines relative to changes they have observed in students' ability to apply the calculus in their fields?

4.   Reform Process

a)   What was the major source of impetus for change? Was the decision made by a small group of faculty or did the total department participate?

b)   Did some of your faculty attend any of the following to learn about calculus reform efforts (if yes, estimate the number of faculty attending)

_____   sessions at the Joint Mathematics Meetings or regional professional meetings.

_____   conferences or workshops sponsored by particular reform projects.

_____   other conferences.

c)   To what degree have reform efforts been adopted and accepted by other faculty members in the department?

d)   Involvement of teaching staff in reform efforts

Group	Much	Some	Little	None
Senior Faculty				
Junior Faculty				
Part-time Faculty				
Graduate TA's				

e)   What hurdles (e.g., faculty resistance, acquiring equipment), if any, did the reform effort have to overcome?

f)   What forms of faculty development were instituted to support the teaching of the reformed calculus (workshop in summer, teaching tips seminar)?

g)   Briefly describe any external funding that you obtained for reform efforts.

h)   Briefly describe how faculty involved in reform calculus teaching have responded to the experience. For example, have the changes demanded more time in preparation, in the development of supplementary materials for class use, in meeting with students individually or in small groups outside of class, in grading, etc.? Do faculty think the additional effort is worth it?

Appendix C
Analysis of Institutional Responses to both Short Surveys

Using the statistical method of capture-recapture, the project reviewed the responses to the 1994 short survey to determine the percentage of institutions (at each degree level) that also returned the 1992 survey. This review showed that 63% of those responding to the 1992 survey also responded to the survey conducted in 1994. For all departments, the response rate in 1994 was about 44%. Table 1 gives an analysis of the response rates to the 1994 survey for each type of institution, comparing all institutions of a given type with the number of institutions that responded to the 1992 survey. It shows that, with the exception of doctoral institutions (for which the 1994 survey had a very high rate of response), the response rate for all institutions of a given type was about equal to or exceeded the response rate for institutions responding to the 1992 survey. This pattern gives support to the proposition that responses to the surveys are not heavily biased towards institutions involved in reform efforts. If one studies the data for, say master's institutions by level of reform, the percentages change little. That is, the percentage of all master's institutions reporting modest reform in the 1994 survey is the same as the percentage of those master's institutions that responded to the 1992 survey. The analysis was limited to the first question on the two mentioned surveys.

The study also indicated that the percentages of 1992 survey respondents who reported different types of change and also responded to the 1994 survey is quite similar to the aggregate information provided in Table 1.

Responses of institutions which answered both short surveys were analyzed for changes in the level of reform. There was a slight difference in the wording of the questions about level of reform. That slight difference in phrasing should not have caused any appreciable differences in the nature of the responses.

Fall 1992:

To what extent has the calculus sequence in your institution changed over the past three years?

  1 no appreciable change   2 modest change
  3 major revision

Spring 1994:

To what extent has the calculus reform movement led to a change in the calculus sequence at your institution over the past three years?

  1 no appreciable change   2 modest change
  3 major revision

Table 1: Recapture Study Degree Level Data

Degree Level	Responses 1994	Total Institutions	Percent 1994	Repeat Responses	Responses 1992	Recapture Percent
Doctoral	150	165	91	58	84	69
Master's	135	236	57	69	117	59
Bachelor's	443	1020	43	222	335	39
Associate	320	1018	31	121	209	22
Total	1048	2439	43	470	745	63

Table 2: Capture-Recapture Institutions by Degree Level

Degree Level	Number Responding
Doctoral	58
Master's	69
Bachelor's	222
Associate	121
Total	470

Table 3: Percentage of Shifts in Response Categories for Institutions in the Recapture Group

From/To	1	2	3	Total
1	20	17	6	43
2	7	29	9	45
3	0	4	8	12
Total	27	59	23	100

A total of 470 institutions responded to both surveys. The distribution of these responding institutions by level of degree program is given in Table 2.

In Table 3, the percentage of the 470 institutions indicating each pattern of shift is displayed. An analysis of the patterns indicates that 57% of the respondents showed no shifts in their programs over the time period between the two surveys. At the same time, 26% indicated a shift up one level and 6% indicated a shift upward of two levels of activity. On the other hand, 11% of the institutions noted a one-category shift backward. These backward shifts may indicate different levels of ac-

tivity in reform at the same institution, ranging from a true movement away from reform to a shift in perception of their level of activity in reform relative to activity at other schools. It may also indicate a change in the perception of the person completing the form.

At each of the degree levels (doctoral, master's, bachelor's, associate), the pattern is one of movement toward reform, but with the majority of schools remaining within the same category of change. For the four levels 59%, 67%, 56% and 57%, respectively, report no change in activity over the two year period. However, 33%, 36%, 32%, and 28%, respectively, report an increase of at least one level of activity over that same time period.

Appendix D
Data on Calculus Enrollments

Information in the following tables was obtained from *Statistical Abstract of Undergraduate Programs in the Mathematical Sciences and Computer Science in the United States (1990–91 CBMS Survey)*, MAA Notes 23, Washington, DC: The Mathematical Association of America, 1992.

Table 1: Calculus I and Calculus II Enrollments (Fall 1990)

Course/Level	Univ (PhD)	Univ (MA)	College (BA)	Two-Year College	TOTAL
Mainstream Calculus I	100,800	38,700	61,800	53,000	254,300
Mainstream Calculus II	47,400	17,200	23,200	23,000	110,800
Non-Mainstream Calculus I	72,700	25,000	50,300	31,000	179,000
Non-Mainstream Calculus II	10,600	2,000	2,000	3,000	17,600
TOTAL	231,500	82,900	137,300	110,000	561,700

(Note: As used here, "mainstream calculus" refers to the course taught for mathematics, physics, and engineering majors. "Non-mainstream calculus" refers to the course taught to biology, behavioral science and business majors.)

Table 2: Full-time Faculty (Fall 1990)

Institution type	No. of Faculty
Univ (PhD)	6427
Univ (MA)	5058
College (BA)	7926
Two-Year College	7222
TOTAL	26,633

Note: The number of part-time faculty has been growing recently. At two-year colleges the number has always been large. The 1990 CBMS survey estimated the number to be 13,680—almost double the number of full-time two-year–college faculty.

Appendix E
Brief Descriptions of Selected Calculus Reform Texts

The Calculus Consortium based at Harvard University

Herman O. Sudholz
Calculus Consortium, Harvard University

The most difficult and most necessary aspect of revitalizing calculus is to get students to think. The traditional calculus has become a litany of procedures and template problems which too often results only in giving students some rather routine practice in algebraic manipulations. In addition, students with weaker backgrounds are usually driven away in frustration over the manipulations required, even if they are able to understand the basic ideas of calculus.

In designing the syllabus, we followed three principles that we believe are important to all calculus renewal projects:

- Start from scratch. Do not look at the old syllabus and try to decide which topics can be left out. It is much better to take a blank piece of paper and decide which topics are so central to calculus that they must be included.

- Show students what calculus can do, not what it can't do. In a beginning college level course, we should be showing students the power of calculus, not the special cases in which it fails. This means, for example, that if we teach Newton's method, we should not emphasize the cases when it doesn't work. We should teach approximations, but not an exhaustive treatment of error estimates. Riemann sums are important, but irregular subdivisions and arbitrary points of evaluation are not. At this stage, complete generality and pathological examples should be kept in the distant background.

- Be realistic about students' abilities and the amount of time they will spend on calculus. In the past, we taught so much so fast that little understanding was developed. It is far better to teach a few topics well.

The Consortium Calculus Course The core calculus curriculum is a three-semester or four-quarter course built around the "Rule of Three." This approach continues the belief that the three aspects of calculus—graphical, numerical, and analytical—should all be emphasized throughout the curriculum. The graphical and numerical aspects of each topic are introduced throughout, along with the analytic. Using a computer or graphing calculator, the extrema of a function can be found by moving the cross-hairs on the graph or by studying the values in a spreadsheet. Equations can be solved numerically and graphically as well as algebraically, which also removes the artificial condition that they all be either quadratic or factorable. Periodic functions make much more sense to students when presented using a graph or table rather than analytically.

The idea of the definite integral comes into the course early in such a way as to make the Fundamental Theorem of Calculus quite transparent. Differential equations are an important part of the course; we find that slope fields provide an excellent way of visualizing the solutions. Tests as well as our presentation must reflect the Rule of Three. In addition, it is important to include non-routine problems on a regular basis. Even if such problems are not difficult or long, the idea that such problems are not only part of mathematics, but, indeed, the point of mathematics, needs to be established.

From 1990 through the spring of 1993, various drafts and the preliminary edition of the single variable text were class-tested. A regular edition of the two semester, single variable text was published in November 1993 along with an Instructors' Manual and various Solutions Manuals. Work is now progressing on a third semester multivariable text, a draft of which is to be available for class testing in fall of 1994. A pre-calculus text is also under development.

The project recommends that faculty using the Consortium materials attend a workshop or minicourse offered through various means. The Consortium sponsors a Conference on the Teaching of Calculus in June and there are workshops and minicourses frequently available at the Joint Mathematics Meetings, meetings of AMATYC, and ICTCM.

57

Calculus (Dick & Patton)
The Oregon State Calculus Connections Project

Thomas Dick
Oregon State University

The text grew out of the NSF-sponsored Oregon State University Calculus Curriculum Project and has been widely disseminated through the Calculus Connections Project.

The text does not differ radically from a traditional calculus text in terms of major topics. Techniques of integration are downplayed. The method of substitution and integration by parts are included in the main text, but other techniques of integration such as partial fractions are relegated to the appendices. Numerical and graphical strategies receive more emphasis.

Making intelligent use of technology   Computer algebra systems, spreadsheets, and graphing calculators are just a few of the readily available technological tools providing students with new windows of understanding and new opportunities for applying calculus. This book seeks to take advantage of these new tools, while at the same time alerting the student to their inherent limitations and the care that must be taken to use technology wisely.

While being "technology-aware," the text itself does not assume the availability of any particular machine or software. Rather, the text adopts a language appropriate for the kinds of numerical, graphical, and symbolic capabilities that are found (and will continue to be found) on a wide variety of computer software packages and sophisticated calculators. For example, the language of "zooming in" on the graph of a function is powerfully suggestive without the need for listing specific keystrokes or syntax. Helping students understand the limitations of technology is a major goal of this text. Students are reminded of the care that must be taken to make intelligent use of computing technology without becoming a victim of its pitfalls. Since no specific hardware or software is assumed, an instructor will need to judge the appropriateness of any particular activity in light of the technology available. However, the exercises are designed to be compatible with a very wide variety of available software and hardware. A graphing calculator will be adequate for most of the activities.

Multiple representation approach to functions   The most important concept in all of mathematics is that of function, and the function concept is central in calculus. The idea of a function as a process accepting inputs and returning outputs can be captured in a variety of representations—numerically as a table of input-output pairs, graphically as a plot of outputs versus inputs, and symbolically as a formula describing or modelling the input-output process. The interpretations of the core calculus topics of limits and continuity, differentiation, and integration all have different flavors when approached through different representations. The text seeks to take a more balanced threefold approach to functions. With each new topic or result, an explicit effort is made to interpret the meaning and consequences in a numerical, graphical, and symbolic context. Such an approach does not require technology, but the availability of an appropriate device allows us greater access to numerical and graphical representations, while at the same time reducing the need for heavy emphasis of rote 'by hand' symbol-manipulation skills.

Visualization and approximation   Two themes that become increasingly important with the availability of technology are visualization and approximation. The ability to obtain a machine-generated graph as a first step instead of a last one can completely turn around our approach to a variety of calculus topics. Graphical interpretation skills become primary. In particular, graphing can be used as a powerful problem solving aid, both in estimating and in monitoring the reasonableness of results obtained numerically or symbolically. Whenever possible, explicit mention is made of the visual interpretation of definitions, theorems, and example solutions, often with direct reference to machine-generated graphs.

Much of calculus grew out of problems of approximation, and many of the key concepts of calculus are best understood as limits of approximations. Numerical tools make once exorbitantly tedious calculations into viable computational estimation strategies. Accordingly, approximation and estimation techniques are given high priority throughout the text.

Calculus from Graphical, Numerical, and Algebraic Points of View

Arnold Ostebee and Paul Zorn
St. Olaf College

Whether one views calculus as an introduction to pure mathematics or as a foundation for applications (or both!), the conclusion is the same—concepts, not techniques, are truly fundamental to the course. Whatever uses they make of the calculus, students need more than a compendium of manipulative techniques. The *sine qua non* for a useful command of the calculus is a conceptual understanding that is deep and flexible enough to accommodate diverse applications.

Our key strategy for improving conceptual understanding is combining, comparing, and moving among graphical, numerical, and algebraic "representations" of central concepts. This strategy pervades and unifies our exposition. By representing and manipulating calculus ideas and objects graphically and numerically, as well as algebraically, we believe that students gain a better, deeper, and more useful understanding.

Combining graphical, numerical, and symbolic viewpoints in calculus can be forbiddingly time-consuming and distracting without technological assistance. With technology, these viewpoints become practically accessible; hence, we use technology freely. However, the text is independent of any particular technology.

Having said this, we emphasize that we do NOT aim to reduce the calculus to mere "button-pushing." We regard computers and graphing calculators strictly as tools albeit very powerful ones—to facilitate the crucial combination of graphical, numerical, and algebraic viewpoints. Our text exploits the capabilities of these tools to help students focus on the ideas that lie at the heart of the calculus.

Our text differs significantly from standard treatments. Graphs or tables of numerical values (or both!) appear on almost every page. For example, in studying differentiation students apply the chain rule to combinations of functions represented in different ways—symbolically, graphically, and by means of tables. Graphical and numerical techniques also complement algebraic approaches to antidifferentiation. For infinite series, routine convergence tests are emphasized less than finding (and defending) numerical limit estimates. Because students repeatedly confront functions that are not defined by algebraic expressions, they must interpret calculus ideas in graphical and numerical terms.

However clear and cogent a textbook's exposition, its problems and exercises generate most of students' mathematical activity and occupy most of their time. Through the work we assign, we tell students concretely what we think they should know and do. Our text includes some problems which are routine drills as well as challenging theoretical problems. It has a greater variety of problems including some posed graphically, some requiring synthesis, and some that ask for comparisons of various solution techniques. As a result, we hope students will see mathematics as an open-ended, creative activity, not a rigid collection of recipes.

Although we emphasize concepts and use technology freely, we by no means ignore symbol manipulations in general, or hand computations in particular. We take the symbolic point of view fully as seriously as the graphical and the numerical.

The text is aimed explicitly at average, mainstream college students (as opposed to honors students or any special disciplinary clientele). To use calculus ideas and techniques successfully, students must know what they are doing and why, not merely *how.* Our text aims at a general audience, including mathematics majors, science and engineering majors, and non-science majors. The content does not depart radically from the norm of most calculus texts. We cover the traditional single-variable topics: function, limit, derivative and antiderivative, integral, differential equations, antidifferentiation, sequences, and series. Basic numerical methods are used to illustrate and reinforce the concepts of integration and convergence. Although our text emphasizes numerical techniques more than usual, we don't "cover" numerical analysis as a topic in its own right. Our choice of numerical methods is governed by how well they illustrate and clarify underlying analytic ideas, not by numerical considerations as such—stability, robustness, etc.

Calculus as A Laboratory Course Project CALC

David A. Smith
Duke University

The Project CALC textbook is called *The Calculus Reader*. The title is a deliberate signal to students that they are expected to read and to learn from reading. Many exercises send the signal that they are also expected to write and to learn from writing. The text covers three semesters of calculus, culminating with vector integral calculus.

Here are some of the characteristics of the text:

1. It has exploration activities embedded in the exposition. Students are expected to be actively involved in explorations with pencil, paper, and graphing calculator.

2. In addition to "drill" exercises (far fewer than traditional texts), it has many conceptual exercises, both within sections and at ends of chapters.

3. The text includes "projects" that are intended to be done by small groups of students over periods ranging from an hour to a week.

4. Interesting problems outside of mathematics are used as motivation to drive the mathematical development.

5. Graphical and numerical approaches are emphasized at least as much as symbolic ones.

6. Inverse problems are stressed. The main theme of the applications is differential equations and first derivatives, then integrals are studied primarily to solve differential equations.

7. The derivative is based on local linearity: Zoom in until the curve becomes straight, and calculate the slope of the line.

8. The integral is based on regular left-hand sums— the Euler's Method solution of an initial value problem that defines the integral. Students are then challenged to find better ways to evaluate integrals.

9. Series are studied primarily as representations of functions. The primary tests of convergence (and estimations of error) are the Alternating Series Test and Comparison to a Geometric Series (the latter leading to the Ratio Test).

10. Some traditional topics (e.g., Integral Test) are treated only in laboratory exercises.

11. The text starts with an extensive treatment of the concept of function—weakest link in students' preparation—with emphases on data, relationships between varying quantities, and graphs. One section in this chapter deals with accuracy of calculations and significance of digits.

12. Antidifferentiation is studied in parallel with differentiation in Calculus I. The definite integral and the Fundamental Theorem appear in Calculus II.

13. The primary motivation in Calculus III is geometry— appropriate after students have had a whole year of intense use of graphics.

14. The text supports an integrated laboratory course for all three semesters. The laboratories are implemented through separate lab manuals using a wide range of technologies, from TI-82's to Mathematica. Users of TI-82, TI-85, or HP-48 versions do not need computer lab facilities until Calculus III. The text assumes the presence of technology and lab activities, but does not mention any particular technology.

15. Theory is downplayed significantly. The words "theorem" and "proof" are seldom used. Students discover many of the truths of calculus and work through plausible reasons that support these truths.

16. Most of the major topics of a traditional course are included, but with significant changes of order and substantially altered emphases. For example, there are no chapters called "Techniques of Differentiation" or "Techniques of Integration." All the important techniques appear in the context of solving recognizably important problems. For example, there is no separate section treating the technique of partial fractions, but the case of two distinct linear factors arises in solving logistic growth problems. In that context it is discussed and treated as the inverse of the problem of adding algebraic fractions.

Calculus in Context
The Five Colleges

Jim Callahan
Smith College

Curricular Goals

1. Develop calculus in the context of scientific questions.

2. Treat systems of differential equations as fundamental objects of study.

3. Construct and analyze mathematical models.

4. Use the method of successive approximations to define and solve problems.

5. Develop geometric visualization with hand-drawn and computer graphics.

6. Give numerical methods a more central role.

Functional Goals

1. Encourage collaborative work.

2. Empower students to use calculus as a language and a tool.

3. Develop students who are comfortable tackling large, messy, ill-defined problems.

4. Foster an experimental attitude towards mathematics.

5. Help students appreciate the value of approximate solutions.

6. Develop the sense that understanding arises out of working on problems, not simply from reading the text and copying its techniques.

Moreover, there are several general principles which shape the organization of the course and the way the concepts of calculus are presented to students.

The Starting Points

1. Calculus is fundamentally a way of dealing with functional relationships that occur in scientific contexts. The techniques of calculus are subordinate to an overall view of the real questions.

2. Computers radically enlarge the range of questions we can explore and the ways we can answer them. Computers are much more than a tool for teaching standard calculus.

3. The concept of a dynamical system is central to science. Therefore, calculus must prepare students to deal with systems of non-linear differential equations.

4. The concept of derivative is much more fundamental than, and is separable from, the process of differentiation.

5. The process of successive approximation is a key tool of calculus, even where the outcome of the process—the limit—cannot be explicitly given in closed form.

These starting points grew out of conversations with scientists and mathematicians, and through our examination of current journals, to see how calculus is actually used. It is typical that scientific problems are poorly posed, noisy, described by incomplete or ambiguous data, and too rich to be solved by a single technique. Mathematics offers a language and a tool for probing the nature of the problem, rather than for solving it in the way students are used to seeing problems solved. There may be a range of acceptable answers rather than one right answer. It was therefore clear to us that we need to foster in our students and experimental attitude towards mathematics. An ability to explore mathematical ideas, to try out a variety of approaches to see what works rather than looking in a book for the "right" technique, is a basic skill for all scientists and mathematicians. In trying to incorporate the implications of these observations into our curriculum, we have found that:

1. The connections between mathematics and the rest of the real world can be made much more immediate.

2. The ability to handle data and perform many computations allows us to explore examples containing more of the messiness of real problems.

3. We can now dcal with models that appear more realistic, and the role of modeling becomes much more central to our subject.

4. Introductory calculus can now offer something substantial to the life and social scientists, as well as to our more traditional clientele in mathematics and the physical sciences.

5. The distinction between pure and applied mathematics becomes even less clear (or useful) than it was.

Calculus&Mathematica

William J. Davis
The Ohio State University, and
Horacio Porta and Jerry Uhl
The University of Illinois

Just as arithmetic is the introduction to the science of counting, calculus is the introduction to the science of measurements—both exact and approximate. *Calculus&Mathematica* presents calculus as a course in exact and approximate measurement. Derivatives measure growth rates, integrals measure accumulated growth, order of contact measures quality of approximation, line integrals measure flow across curves, the divergence measures the strength of a source or sink. We want the student to think of calculus as a toolbox of measurement devices. *Calculus&Mathematica* consists of learning what the tools are and how to use them for systematic scientific measurements both exact and approximate.

The 1989 National Research Council report *Everybody Counts* points out: "Scientific computation has become so much a part of everyday experience of scientific and engineering practice that it can be considered a third fundamental methodology of science—parallel to the more established paradigms of experimental and theoretical science." *Calculus&Mathematica* was designed with this in mind. Once you take this point of view, then you can see why many well-accepted calculus rituals must be scrapped and replaced by worthier activities. The biggest change this forces is acceptance of the a priori premise that functions have instantaneous growth rates (slopes), regions have area, curves have length, etc. The job of calculus is to measure these and other quantities by use of functions, derivatives, (definite) integrals, and approximations.

Calculus&Mathematica attempts to involve students actively in their own learning by putting them in a position to acquire mathematical ideas visually in their own way and at their own pace. Instead of hanging technology on the end of the learning process, *Calculus&Mathematica* uses technology to initiate the learning process. It accomplishes this through its electronic interactive text, structured around notebooks, and it invites students to write about the mathematics they are doing as they are doing it.

Each notebook begins with basic problems introducing the new ideas, followed by tutorial problems on techniques and applications. Both problem sets have "electronically active" solutions to support student learning. The notebook closes with a section called "Give-it-a-Try," where no solutions are given.

Students use both the built-in word processor and the graphic and calculating software to build their own notebooks to solve these problems, which are submitted electronically for comments and grading.

Notebooks have the versatility to allow reworking of examples with different numbers and functions, to provide for the insertion of commentary to explain concepts, to incorporate graphs, and plots as desired by students, and to launch routines that extend the complexity of the problem. The instructional focus is on the computer laboratory and the electronic notebook, with less than one hour per week spent in the classroom. Students spend more time than in a traditional course and arrive at a better understanding, since they have the freedom to investigate, rethink, redo and adapt. Moreover, creating course notebooks strengthens students' sense of accomplishment.

The tone of *Calculus&Mathematica* is not "mathematical" in the sense that most students would use this word. On the contrary, the tone is nearly always conversational, sometimes humorous and occasionally a bit irreverent. It is written this way to let the students know that the course is on their side in the learning process. The style is good enough to get the mathematics ideas across and it is informal enough for the students to understand and for the students to imitate in their own writing.

Aspects of the traditional course deemphasized (but not necessarily absent) in *Calculus&Mathematica*:

- Algebra as the primary delivery vehicle of calculus
- Polynomials
- Tangent lines
- The "indefinite integral" without limits
- Right triangle trigonometry
- L'Hospital's Rule
- Infinite sequences of numbers and infinite series of numbers
- Error terms for truncated Taylor series
- The litany of convergence tests
- Convergence at the endpoints.

Calculus&Mathematica opens with two lessons on how the two most important calculus functions—the line functions and the exponential functions—are used in the real world. Trigonometric functions come in early too, but they are studied for their oscillatory properties rather than their definitions in terms of triangles. Other functions enter the course on an as needed basis.

Several professors from diverse fields of science and engineering have pointed out to us that polynomials simply do not come up much outside the traditional calculus classroom. And they all went on to say that the polynomial calculus studied so carefully in the traditional course does not make a successful transition to their classes. Their view is that the functions most deserving of calculus study generally involve exponential functions and trigonometric functions and combinations of them. We have taken their advice very seriously. This is why exponential and trigonometric functions are in the hopper in the very first lesson on instantaneous growth.

C⁴L: Calculus, Concepts, Computers and Cooperative Learning (Dubinsky, Mathews, and Schwingendorf)

The Purdue Calculus Reform Project
David Mathews
Longwood College

The Purdue Calculus Reform Project, C⁴L (Calculus, Concepts, Computers, and Cooperative Learning) is part of a comprehensive curriculum reform program concerned with developing new courses in precalculus, calculus, discrete mathematics, and abstract algebra. The project is also beginning to work on other courses such as linear algebra.

The project attempts a comprehensive reform of each course, emphasizing technology and the development of pedagogical strategies that resonate with what we are learning from research about how people learn mathematics. We have a particular emphasis on cooperative learning and having students write programs as a means of constructing mathematical concepts in their own minds.

The textbooks *Calculus, Concepts and Computers (Parts I and II)* and *Calculus, Concepts and Computers for Management, the Social and Life Sciences* of the C⁴L Project support a constructivist approach to teaching. That is, they can be used in undergraduate calculus courses to help create an environment in which students construct for themselves mathematical concepts appropriate to understanding and solving problems in calculus.

The texts are divided into sections, each intended to be covered by an average class in about a week. Each section consists of a set of activities, class discussion material, and a set of exercises, which forms the basis of our Activity-Class-Exercise (or ACE) Cycle.

The Activities of the text require students to write short computer code in a mathematical programming language, like ISETL or parts of MAPLE V, to represent mathematical concepts. Often, an activity will require use of mathematical ideas not yet covered in the text. The student is expected to try to discover the mathematics or sometimes just make guesses, possibly reading ahead in the text for clues or explanations. As a result of making mathematical constructions themselves on a computer during this Activity phase, students in turn make the mental constructions necessary for understanding calculus.

The Class material contains some explanations, some completed mathematics, and many questions, all taking place under the assumption that the student has already spent considerable time and effort on the Activities related to the same topics. Our experience indicates that with the foundation established through working with the Activities, students relate in a more meaningful way to formal definitions and theorems. Each unanswered question in the text is either answered in a later section or repeated as an explicit problem in the exercises. Students working together in teams on the discussion materials largely replaces lectures. Lectures are used to summarize learning after the students have had a chance, through the activities and discussions, to understand the material.

Finally, the Exercises in the texts are relatively traditional and are used to reinforce the ideas that the students have constructed through the Activity and Class phases. Though the teaching method supported by this approach is novel, the selection of material is fairly standard. The texts contain all of the material usually covered in standard calculus courses on single and multivariable calculus together with an introduction to vector calculus. The first chapter of *Calculus, Concepts and Computers* (Part I) provides the reader with the opportunity to practice with a mathematical programming language and a symbolic computer system, both of which will be used throughout the book. In addition, Chapter 1 introduces the basic problems and ideas of both differential and integral calculus, or rather, "approximate calculus" and points toward the notion of limit. Chapter 2 refines the concept of function and attempts to deepen student understanding of this concept so necessary for the study of calculus and other areas of mathematics. From our recent research, and that of others, on how students learn calculus, we believe that the time spent on functions and introducing the basic problems and ideas of calculus is well worth the effort. Chapter 3 covers the concept of the limit of a function at a point together with the notions of limits at infinity and infinite limits. In Chapter 4 we cover the basic concepts and related problem solving situations in differential calculus. In Chapter 5 we cover the concepts of integral calculus together with elementary differential equations. Chapter 6 deals with sequences and series of numbers together with power series and Taylor series representations of functions. Chapter 7 concludes Part I with polar coordinates, vectors, parametric equations, and curvilinear motion along with vector-valued functions.

Part II of *Calculus, Concepts and Computers* begins with Chapter 8 where the ideas of differentiation from Chapter 4 are extended to functions of more than one variable. Chapter 9 extends the ideas of integration in Chapter 5 to functions of more than one variable. Finally, Chapter 10 provides a brief introduction to the ideas of vector calculus, culminating with the famous theorems of Green, Stokes and Gauss. Our text *Calculus, Concepts and Computers for Management, the Social and Life Sciences* provides an analogous development of single and multi-variable calculus and applications for this different audience.

One feature of our pedagogical approach which is apparent in all of the texts is a "holistic spray" of ideas of calculus, in a spiraling manner, through the activities and text discussion. This is done to provide students with ample opportunity to do the necessary mental constructions needed for a deeper understanding of the basic concepts. The texts are intended to be used in a cooperative (small group) learning environment to foster reflective thinking. An extensive *Instructor's Resource Package* is available for each of the texts.

Calculus Using Mathematica
The University of Iowa

Keith D. Stroyan
The University of Iowa

The University of Iowa offers a special new calculus course. In addition to developing students' basic skills and concepts, the course presents calculus as the "language of science" and has students actively doing mathematics in laboratory-type projects. The dedicated labs associated with the course use Mathematica computing for once-a-week electronic homework as well as in the assigned larger projects of the course.

Students develop skills and then learn why calculus is important by applying their skills to answer difficult questions of clear contemporary interest such as: (1) What is a sustainable harvest level for Sei Whales? (2) Why was polio eradicated by vaccination, but not measles? (3) Will a bungee diver smash into the bottom of the canyon? (4) How do you calculate the inverse of $y = x^x$ yourself? (5) What is the resonant frequency of an electrical circuit? Carefully selected scientific problems guide the curriculum and motivate the ideas of calculus for students. Their answers to these large problems are presented in the form of 10 (or so)-page term papers once or twice per semester.

Mathematica programs (called NoteBooks) play several important roles in this presentation of calculus. First, computer use illustrates and reinforces basic mathematical concepts. Second, up-to-date scientific computing helps to reduce technicalities and concentrate on the main ideas of calculus. Access to computing also allows the use of more realistic applications that are too "messy" to compute by hand.

The course is accelerated in that the main topics from roughly three semesters of traditional calculus are covered in one year. This is a rough comparison, because modern computing allows us to treat "advanced" topics like phase plane analysis of nonlinear differential equations as well. Engineering students who successfully work the project on resonance are exempted from sophomore engineering differential equations.

You will notice that the table of contents does not look like the one in a regular calculus book. Also, the list of projects is large so students can select topics of interest. We do not expect any student to work more than four major projects per year.

In Accelerated Calculus, we complete all the topics in the core text in one year, but we also select students who are very well prepared in high school mathematics. (Many have had warm-up high school calculus, too.)

Since we have about 1,000 calculus students at Iowa, this selection makes sense to us. About three fourths of this material is learned by ordinary students in a year, simply by spending more time on drill material. Perhaps with the addition of a chapter on surface and volume integrals, the text could serve average students for three full semesters and include all the basic skills of the traditional three-semester course as well as the new topics on discrete and continuous dynamical systems.

Because of the computer lab, our expansion plans depend on obtaining new equipment, but our program is so successful that we want to move ahead with regular calculus as fast as we can install the needed equipment. The initial cost of equipment is high and the problem of finding space is difficult to solve, but the long-term cost per student is close to our new computing fee of $40. We feel that this is a modest cost for a modern lab course.

Projects are tremendously valuable. Students become very excited and involved in them. Faculty members have walked past our lab and asked me, 'What are you doing in there? I've never seen that much excitement in a math course.' Projects give students a new view of mathematics which involves thinking about what it means. However, student technical writing is often very hard to evaluate. Occasionally there is a lack of cooperation between project team members, but the benefits of team interaction far outweigh the problems. Often teams do not take full advantage of the first submission of their reports. First submission needs to carry credit and in addition we have tried brief oral presentations early on. This helps get people started. Our students are very diverse, ranging from music to engineering, so we have many choices of project topics. Hence, there is some additional value to having the whole class hear a description of the other projects. Often the mathematical heart of a project is one important mathematical idea. Two weeks of work on and off usually lead the student to clear understanding of that idea. Experience is the best teacher. It ought to be; think how much it costs and how long it takes. Our approach is a combination of 'force feeding,' skill development, difficult problems, and term paper projects, because none of these methods is perfect by itself.

Much of the presentation is in the traditional style. "Reformers" may not approve, but we feel that an optimum course will use lecture/discussion and many traditional topics. The foregoing makes it clear that we also believe the optimum course will use technology creatively and have students deeply and actively involved in doing and using calculus a substantial part of the time. We are still working on the right balance between these things. Just trying to strike this balance makes the course worthwhile for students.

Appendix F

List of Awards
in the
NSF Calculus Program

PROJECT ABSTRACTS: FY 1988 AWARDS

An Integrated Program in Calculus and Physics

F. Richard Yeatts
Colorado School of Mines
Golden, CO 80401

Award No: USE 8813784
FY 88 $ 74,517
FY 89 $ 47,451
FY 90 $ 83,058

An integrated calculus and physics course is being developed and tested. A unique feature is the well-planned laboratory/workshop session where much of the integration of the subject matter occurs. The sessions provide students with the opportunity to explore and discover the relationships between a physical situation, its graphical or geometric representation, and the corresponding analytical representation. The workshop exercises consist of physics experiments, numerical simulations, symbolic manipulations, computer programming, and formal reasoning exercises. Study guides, problem sets, and modular materials are being developed. The students' progress is being evaluated internally as well as externally by a National Advisory Committee.

The Design of a Computer Algebra System to Effect a More Relevant Mathematics Curriculum

J. Douglas Child
Rollins College
Winter Park, FL 32789

Award No: USE 8814048
FY 88 $ 73,436
FY 89 $ 43,126
FY 90 $ 44,076

The focus of the three-year project is the construction of a computer environment consisting of a computer algebra system, MAPLE, a specially designed interface to MAPLE, a hypertext system, and other software that is more suitable for teaching and learning calculus for the average student. The computer algebra system demonstrates the reasoning processes of experts. The intent is that students will learn how to think about solving calculus as well as how to solve problems with the help of a computer algebra system. The computer environment is suitable for pre-calculus, science, and engineering curriculum designs. A computer algebra system, both to do and to teach calculus with MAPLE, is being developed along with curriculum and interfaces for computer algebra systems. The calculus topics are being reordered introducing differentiation and integration early in the course via applied problems. Emphasis is being placed on logic, precise use of language, numerical methods, approximations, and mathematical modeling. Experimental use of materials is taking place at colleges and local high schools which have classes of approximately thirty-five students. National dissemination is

in the form of text to be published by Wadsworth/Brooks-Cole Publishers.

Calculus Workshops and Conferences

Shair Ahmad
University of Miami
Coral Gables, FL 33124

Award No: USE 8813860
FY 88 $ 45,000
12 months

The project is developing a monthly series of two-day conferences and workshops on calculus to be attended by university and community college faculty members, high school calculus instructors, and industrial representatives. The seminars concentrate on the role of computers and calculators, textbooks, relevance to other disciplines, conceptual understanding, and the development of exercises that stress current technology. Discussions are led by small groups of well-prepared individuals familiar with existing literature on the subjects. Participants are encouraged to carry on similar discussions in their own institutions.

Calculus Planning Project

Nagambal D. Shah
Spelman College
Atlanta, GA 30314

Award No: USE 8813792
FY 88 $ 50,000
12 months

The project is planning a series of seminars focusing on the special needs that women have in the study of calculus and mathematics, and culminating in a faculty retreat and the writing of a report on the faculty members shared experiences. The project is being held at a black institution for females with strong academic traditions and includes formal participation by the chairman of the mathematics department at Agnes Scott College, a female institution with equally strong academic traditions. The program starts with three seminars lead by consultants with special expertise in the area of female studies in mathematics. The presentations and consultations should sensitize the faculty participants to the special elements of females studying mathematics. Three additional consultants will assist with three later seminars that concentrate on the role of computers in the study of calculus. A pilot section of calculus uses and investigates computer software specially designed to assist in the study of calculus. Three undergraduate students help with the evaluation of the materials. With the help of the specialized consultants, the faculty becomes sensitive to the issues involved in the instruction of female student.

Calculus Curriculum Development

Gerald J. Janusz   Award No: USE 8813873
University of Illinois   FY 88 $ 40,785
Urbana, IL 61801   12 months

Mathematicians are investigating methods by which the teaching of calculus can be made more effective in conveying to students an understanding of calculus as a powerful problem solving tool. Course material planning focuses on the development of the problem sets that lead students through the central ideas and methods of calculus and enhance their ability to read and write mathematics. The project consults with user departments in science, engineering, and other areas; develops and tests course material for Calculus I based on an approach of Artin; focuses on computational and problem solving; and develop student capacity to read and write mathematics correctly and coherently. A weekly Calculus Seminar trains teaching assistants and discusses content among faculty within and outside the mathematics department. A Calculus Workshop for faculty and high school teachers is being held.

Calculus, Concepts and Computers

Edward L. Dubinsky   Award No: USE 8813996
Purdue University   FY 88 $ 30,000
West Lafayette, IN 47907   12 months

Three mathematicians are teaching three small (25-30 students) prototype calculus courses based on computer, computer languages, and algebra systems. The geometric and conceptual aspects of calculus, solution of applied problems, and reduction of routine drill by using symbolic manipulation is emphasized. These courses include both mainstream and non-mainstream calculus. The extent of the use of computer labs and the use of the ISTEL and Maple software packages varies. A consulting board of 25 experts from the various academic disciplines will suggest applications from science and mathematics from the latter third of the 20th century. A unique component of the project is research on how students come to understand the underlying ideas in calculus. Theoretical analysis, observations, and experiments on the teaching and learning of calculus are being formulated.

Planning for a Revitalization of an Engineering/ Physical Science Calculus

Elgin H. Johnston   Award No: USE 8813895
Iowa State University   FY 88 $ 49,954
Ames, IA 50011   12 months

Mathematicians are revising the engineering calculus sequence by incorporating modeling and symbolic/graphical/ numerical software into the curriculum. The planning is done by a committee of faculty from engineering, physical, and mathematical sciences. A calculus network of high school, community college, and college/university is being established. Over 90% of the calculus students are from the client disciplines. The project initiates change in a deliberate and timely manner with concurrence by the client departments. One fourth of the first-semester calculus courses are taught under the revised curriculum requiring a small amount of programming, and stressing algorithms and sharply focused real applications. The second and third semesters of calculus are being revised within the same mathematical framework.

Dynamic Calculus

Robert L. Devaney   Award No: USE 8813865
Boston University   FY 88 $ 40,306
Boston, MA 02215   12 months

An expert fluid dynamicist is developing instructional modules, which incorporate ideas from modern dynamical systems theory into the standard introductory calculus course. The purpose of the project is to augment the calculus with topics of current research interest. Materials can be introduced early in a calculus sequence so that students receive early exposure to topics of contemporary research interest in mathematics, computer experimentation in mathematics, and exciting mathematical visual images. Several modules which show how certain topics in dynamics may be integrated into calculus, and the role of dynamical calculus in science are being developed.

Calculus in Context

James Callahan   Award No: USE 8814004
Five Colleges, Inc.   FY 88 $ 141,707
Amherst, MA 01002   FY 89 $ 190,845
                        FY 90 $ 174,183
                        FY 91 $ 129,392
                        FY 92 $ 74,128

Mathematicians are restructuring the standard three-semester calculus sequence. A new curriculum is being developed in which the four mathematical themes of optimization, estimation and approximation, differential equations, and functions of several variables are stressed from the beginning. These major mathematical concepts grow out of exploring significant problems from social, life, and physical sciences. Dynamical systems, discrete time models, Fourier series, and partial differential equations are some of the concepts which are explored. The computer is being integrated into the curriculum as a basic conceptual device for structuring the way students think about problems and what it means to solve them. Dissemination is in the form of team-taught courses, weekend retreats, summer workshops for area faculty and high school teachers, and publication of the curriculum. These instructional materials are used at universities, liberal arts colleges, and high schools.

The Language of Change: A Project to Rejuvenate Calculus Instruction

Andrew M. Gleason	Award No: USE 8813997
Harvard University	FY 88 $ 20,362
Cambridge, MA 02138	12 months

A group of mathematicians is designing calculus syllabi outlines. They are investigating the use of computers/calculators in opening up new topics and new ways of teaching. They are completely rethinking the goals and content of calculus courses to establish closer collaboration with representative of client disciplines; to plan the creation of tests; and to plan the development of materials to be used in workshops on pedagogy.

Calculus Reform in Liberal Arts College

A. Wayne Roberts	Award No: USE 8813914
Macalester College	FY 88 $ 62,650
Saint Paul, MN 55105	12 months

Mathematicians are developing a one-year mainstream calculus course. The curriculum stresses basic concepts; numeric and graphic experiments to better understand the power and limitations of technology; the role that calculus plays in changing people's world view; the art of writing a deductive argument; and applied mathematics as a creative modeling process. Outlines of teaching resources to create a lean and lively one-year calculus course include sequences of laboratory style problems; textbook type problems for computers/calculators; application modules; open ended problems; and historical vignettes.

Calculus: Restructuring and Integration with Computing

Richard H. Crowell	Award No: USE 8814009
Dartmouth College	FY 88 $ 50,464
Hanover, NH 03755	12 months

Mathematicians are integrating computers into, and planning the restructuring of, the calculus curriculum. The approach emphasizes elementary functions and the use of the computer for graphical displays and computation of tables of function values. Student written programs are being used to investigate these functions. Differentiation and integration are taught by means of the difference cal-

culus, making heavy use of the computer. These concepts are used to solve "real-world" problems. Students are expected to gain a deeper understanding of calculus concepts from the combination of theory, applications, and computer investigations. During the four-course calculus sequence, students are developing the capabilities of doing their own numerical and graphical investigations independently. Text and computer materials, demonstration programs, and problems are being developed. The new curriculum is being tested and evaluated.

From Euclid to von Neumann, an Activity-Based Learning Experience in Calculus: Project ENABLE

Joan Ferrini-Mundy	Award No: USE 8814057
University of New Hampshire	FY 88 $ 40,487
Durham, NH 03824	12 months

Mathematicians, mathematics educators, engineers, scientists, high school teachers, and Technical Education Research Center are developing and refining mathematical, educational, and technological perspectives for a three-semester calculus curriculum. The project is first conducting a baseline assessment of first-semester calculus students to determine their algebra and trigonometry skills, as well as their understanding of essential precalculus concepts. Implementation of the reorganized and streamline curriculum requires a clear perception of the students' knowledge base and misconceptions, as well as the students' active participation in their own learning. The curriculum starts with the concept of approximation, whose idealization will lead to derivatives, integrals, and continuity. Biweekly seminars to develop prototype materials, outline modules, and core units are being held during the academic year. Some components are being tested in high schools.

Student Research Projects in the Calculus Curriculum

Marcus S. Cohen	Award No: USE 8813904
New Mexico State University	FY 88 $ 83,572
Las Cruces, NM 88003	FY 89 $ 98,993
	FY 90 $ 59,652

Mathematicians continue to develop and implement a plan using student research projects in a broad range of calculus courses. Three individual two-week projects are being used instead of hour exams in thirty-five calculus sections. Almost one-half of the students are minorities. A collection of four hundred problems, many annotated with information on their success in the classroom is being compiled. The projects require that students think broadly and deeply, identifying background material, and synthesizing an approach. Scientists, engineers, and economists help design projects which demonstrate the mathematical underpinnings of solutions to applied problems. Faculty workshops, training of teaching graduate assistants, Advisory Committee meetings, and an extensive evaluation which includes evaluating long-term intellectual growth of students are all taking place as part of the project.

Planning a Problems-Based Calculus Curriculum

Stephen R. Hilbert   Award No: USE 8814177
Ithaca College   FY 88 $ 50,193
Ithaca, NY 14850   18 months

Mathematicians are developing a problem-based mainstream calculus curriculum. The complex problems will require a minimum of several weeks to solve. The structure of the problems varies from "case study" to "open-ended". Groups of students working together will solve problems which develop essential parts of the calculus, and use calculator/computers where relevant. In-depth interviews with twenty-five faculty members from accounting, biology, chemistry, economics, finance, management, politics, psychology, and physics help identify realistic problems. Two experimental course sections of Calculus I are being taught, and a one-day conference on the "Future of Calculus" is being held.

Calculus and the Computer: Innovative Teaching and Learning

William E. Boyce   Award No: USE 8814011
Rensselaer Polytechnic Institute   FY 88 $ 50,000
Troy, NY 12180-3590   12 months

Applied mathematicians are developing a calculus course sequence in which computer technology is used to equip students with powerful and versatile problem-solving tools in order to gain deeper understanding of the underlying mathematical concepts. Content includes numerical computation, sophisticated graphics, symbolic computation, relations between mathematics and the natural world, and mathematical modeling. A team of mathematicians, a physicist, an electrical engineer, and students are developing instructional materials to support the use of the computer in calculus. These materials are used in two or three pilot sections of the calculus sequence.

Development of Calculus

Lawrence C. Moore   Award No: USE 8814083
Duke University   FY 88 $ 20,000
Durham, NC 27706   12 months

A detailed syllabus for a new calculus curriculum is being developed in a cooperative venture with an area high school. The schools are experimenting with the use of computer algebra systems and preparing sample modules. A small prototype calculus laboratory is being run for the development and testing of interactive experiments and writing as a learning tool in mathematics. Central themes include appropriate use of mathematical and physical tools, identification of a concept and its inverse, use of transformations, and relationships between calculus and real world problems.

Toward a Conceptual and Captivating Calculus

Thomas A. Farmer   Award No: USE 8813786
Miami University Oxford Campus   FY 88 $ 48,595
Oxford, OH 45056   12 months

The project is developing a lean and lively calculus syllabus for college students who have had calculus in high school. After consultation with scientists from client disciplines on the current uses of calculus, materials incorporating computers are being developed to run a preliminary experiment. A large-scale, controlled experiment with teaching and computing materials is being prepared.

Plan for Calculators in the Calculus Curriculum

Thomas Dick   Award No: USE 8813785
Oregon State University   FY 88 $ 27,401
Corvallis, OR 97331-5503   12 months

A calculus curriculum which makes essential use of the HP285 symbolic/graphical calculator is being developed and implemented. The objectives are: to identify calculus topics pedagogically suited for use on symbolic/graphical calculators; to identify roles of symbolic/graphical calculators in calculus; and the production of curriculum/calculator materials to be tested during the academic year. A calculus book based on the symbolic/graphical calculator provides both technical advice regarding the calculator and adaptation of the calculus text materials. A series of workshops on utilizing the symbolic/graphical calculator in mathematics classes is being presented.

Integrated Calculus Development

Alain Schremmer   Award No: USE 8814000
Community College of Philadelphia   FY 88 $ 40,124
Philadelphia, PA 19107   12 months

The project is developing a Lagrangian calculus program for the students who are predominantly women, minorities, and returning adults. Lagrangian calculus develops concepts via polynomial approximations rather than limits. It reduces questions about "any" function to the same question about a power function, which appears in the approximating polynomial. The two-semester course is equivalent to pre-calculus and one semester of calculus. The first semester consists of linear approximations, quadratic, and power functions. The second semester consists of the differential study of polynomials, Laurent polynomials, and rational and elementary transcendental functions by Lagrange's approach.

Revitalization of Calculus

Mary McCammon　　　　　　Award No: USE 8813779
Pennsylvania State University　　　　FY 88 $ 42,399
University Park, PA 16801　　　　　　12 months

Mathematics faculty members are developing a lean and lively syllabus for a freshman science and engineering calculus sequence. In consultation with other mathematicians, scientists, and engineers, a core of essential material is being determined. Existing software, computer technology, and placement tests are evaluated and modified, as needed. A test which contains related software and supplements for instructors and which covers the central core of materials is being produced. Several aspects of the experimental syllabus are being taught by the instructors. Each participant experiments with only a small part of the curriculum. In this way, several content areas and approaches can be tested, while insuring that students are exposed to nearly all of the traditional calculus.

Proposal for a Newsletter on Collegiate Mathematics Education

James H. Voytuk　　　　　　Award No: USE 8814683
American Mathematical Society　　　　FY 88 $ 104,413
Providence, RI 02901　　　　　　　FY 89 $　66,675
　　　　　　　　　　　　　　　　FY 90 $　　9,358
　　　　　　　　　　　　　　　　FY 91 $　　9,358

A collegiate mathematics education newsletter is being established. Its purpose is to stimulate greater communication between research mathematicians and collegiate mathematics educators. The newsletter provides a balance of short, timely items directing readers to sources of further information, and longer, more substantive articles presenting discussion of important issues in collegiate mathematics education. The newsletter includes the following: articles on mathematics curriculum; innovative teaching methods; funding for collegiate mathematics education; outside classroom activities; profiles of successful mathematics programs; information on conferences, workshops, courses, and use of technology; review of international activities; review of information in other publications; and a column for queries.

The Calculus Companion: A Computerized Tutor and Computational Aid

Edmund A. Lamagna　　　　　Award No: USE 8814017
University of Rhode Island　　　　　FY 88 $ 51,350
Kingston, RI 02881　　　　　　　　12 months

The project is creating a computational environment in which calculus students use the computer as both a tutoring device and a computational aid. The system consists of two components: (1) a powerful user interface to a symbolic mathematics package and graphical display routines; and (2) tutorial modules. A study on how the computer can be best integrated into the calculus curriculum, and a prototypical course module on the topic of integration are being completed. Students are introduced to important techniques using symbolic computation facilities. The graphical and numerical capabilities demonstrate several numerical integration techniques. Real world examples from several client disciplines are used to motivate topics.

Restructuring One Variable Calculus within a Modeling and Computer Oriented Environment

Daniel C. Sloughter　　　　　Award No: USE 8813781
Furman University　　　　　　　　FY 88 $ 22,476
Greenville, SC 29613　　　　　　　12 months

Mathematicians are developing and testing an experimental one variable calculus course which builds and analyzes realistic models of dynamic processes, including "chaos". The course restructuring starts with sequences of real numbers and difference equations, and ends with differential equations. Global and qualitative behavior is stressed by use of the computer. There is modeling with symbol manipulation, discrete mathematics, and numerical mathematical packages.

Development of Computer-Based Curriculum Materials for Calculus: A Planning Project

Michael E. Moody　　　　　　Award No: USE 8814131
Washington State University　　　　FY 88 $ 29,716
Pullman, WA 99164-3140　　　　　　12 months

Mathematicians are developing, coordinating, and writing multi-disciplined computer-based curriculum materials for calculus to be implemented at two high schools, a community college, a private college, and a public university. Generic curriculum materials for engineering calculus, calculus for life sciences, and business calculus are being developed by faculty from engineering, biology, chemistry, business, and sociology. The materials include laboratory exercises that use computing devices such as HP-28 and microcomputers using symbolic manipulation programs. These realistic problems use numerical methods that illustrate the power, difficulties, and logic of computation and graphical solution to problems. Electronic "slide shows" with both animated and static computer graphics of classroom demonstrations and lectures are also being developed.

PROJECT ABSTRACTS: FY 1989 AWARDS

Curriculum Development Project: Calculus

David O. Lomen
University of Arizona
Tucson, AZ 85721

Award No: USE 8953930
FY 89 $ 104,806
12 months

Materials are being developed that complement the calculus courses at major universities throughout the nation. Integrated supplements are being developed that feature laboratories, projects, problems, and software packages. Laboratories are modeled after a typical physics or chemistry laboratory where the student performs guided experiments independent of the present class material. Projects involve the student discovering and conjecturing results related to calculus. Problems are challenging, realistic questions that might require modern technology to solve. All problems are technology dependent, but independent of a specific brand of computer. The software packages will bridge this gap by supplying the appropriate materials for MS-DOS and Macintosh machines.

Calculus and Computers: Toward a Curriculum for the 1990s

Marcia C. Linn
University of California
Berkeley, CA 94720

Award No: USE 8953974
FY 89 $ 42,898
12 months

Faculty from a broad spectrum of institutions, including two-year colleges, are learning ways to use Mathematics and exchanging ideas on how to use this powerful tool in the teaching of calculus. The invited speakers at the conference and the PIs are using Mathematica and other integrated symbol manipulation and graphics systems in their calculus courses and are seeing exciting possibilities for their use. The conference participants are learning about these systems and are making suggestions about ways to use these tools.

Rapid Dissemination of New Calculus Projects

Thomas W. Tucker
Mathematical Association of America
Washington, DC 20001-0000

Award No: USE 8953912
FY 89 $ 41,540
12 months

Detailed descriptions (syllabi, assignments, laboratories, exams, sample text material, preliminary assessment) of eight to ten new calculus projects are being prepared for publication. Project summaries of approximately 50 additional projects are included.

Calculus Curriculum Development

J. Jerry Uhl
University of Illinois
Urbana, IL 61801

Award No: USE 8953906
FY 89 $ 25,000
12 months

A non-traditional, entirely new course is being developed under this pilot project through live Mathematica notebooks. Emphasis is placed upon individual student use of the Mathematica program for instruction, computation, and symbolic manipulation within the Mathematica notebooks. The goal is to motivate the students to better understand the foundations and enable them to execute calculations far beyond those expected of students in the traditional course.

Calculus Redux

Judith H. Morrel
Butler University
Indianapolis, IN 46208

Award No: USE 8953948
FY 89 $ 27,000
12 months

Students are finding more excitement and making better progress in calculus because of a revised curriculum that emphasizes problem solving, building intuition, and improving written mathematical expression. A data base consisting of non-routine, open-ended, multi-step problems and discussion modules emphasizing concepts, experimentation, and widely varying applications is being created.

A Revitalization of an Engineering/Physical Science Calculus

Elgin H. Johnston
Iowa State University
Ames, IA 50011

Award No: USE 8953949
FY 89 $ 63,600
FY 90 $ 72,250
FY 91 $ 58,565
FY 92 $ 15,000

A four-year program is under way to revitalize the calculus course taken by science, engineering, and mathematics students. The revised curriculum stresses the modeling and problem-solving aspects of calculus, and teaches students to use commercially available symbolic and numerical software to handle the technical aspects of the subject. The planning, testing, and implementation of the new curriculum is being done under the guidance of a liaison committee made up of faculty from the physical sciences, engineering, and mathematics departments.

Calculus with Computing: A National Model Course

Keith D. Stroyan	Award No: USE 8953937
University of Iowa	FY 89 $ 65,000
Iowa City, IA 52242	12 months

The curriculum is being developed to present calculus as the language of science. Beginning calculus is being treated as a laboratory course with modern computers and scientific software as the laboratory equipment. The development is built on a long history of successful use of computers in a calculus laboratory and will make use of new software so that students have a serious start on their education in scientific computation.

Core Calculus Consortium: A Nationwide Project

Andrew M. Gleason	Award No: USE 8953923
Harvard University	FY 89 $ 346,245
Cambridge, MA 02138	FY 90 $ 570,283
	FY 91 $ 335,223
	FY 92 $ 418,372
	FY 93 $ 337,500

A National consortium of institutions is developing an innovative core calculus curriculum that is practical and attractive to a multitude of institutions. The refocus of calculus uses the "Rule of Three" whereby topics are explored graphically, numerically, and analytically. The consortium is led by Harvard University and consists of the University of Arizona, Colgate University, Haverford-Bryn Mawr Colleges, the University of Southern Mississippi, Stanford University, Suffolk Community College, and Chelmsford High School.

Calculus Reform in Liberal Arts College

A. Wayne Roberts	Award No: USE 8953947
Macalester College	FY 89 $ 199,203
Saint Paul, MN 55105	FY 90 $ 215,168
	FY 91 $ 148,500

A calculus curriculum is being developed that stresses understanding rather than techniques, contains realistic applications, and promotes the ability to write coherent arguments. This development is being carried out with the participation of twenty-six liberal arts colleges in the Midwest and takes the form of five Resource Collections containing fundamental materials that can be used in part or in total for curriculum development in calculus at any institution. These collections are to be published as five separate volumes.

The St. Olaf Conference, October 20-22, 1989

Paul D. Humke	Award No: USE 8955091
Saint Olaf College	FY 89 $ 1,500
Northfield, MN 55057	6 months

Mathematicians experienced in using computer algebra systems in teaching calculus are meeting to discuss their past experience and their plans for future use. The focus is on how these systems have changed, can change, and will change the teaching of calculus.

Utilization of Technology in Non-traditional Calculus

Wanda Dixon	Award No: USE 8953931
Meridian Community College	FY 89 $ 25,000
Meridian, MS 39301	18 months

The calculus curriculum is being revised to place more emphasis on learning the concepts, solving realistic problems, and improving estimation of skills. Materials are being developed to utilize the HP-285 hand-held calculator.

Calculus: Restructuring and Integration with Computing

Richard H. Crowell	Award No: USE 8953908
Dartmouth College	FY 89 $ 289,171
Hanover, NH 03755	

The calculus curriculum is being restructured by integrating into the syllabus student use of a personal computer as a working tool. A substantial body of new courseware is being created that enables the students to use a personal computer as a regular part of their homework to explore, analyze, or verify the central concepts of the calculus, is being created. Students are expected to write some of their own software and/or to modify existing software as an integral part of the course. The course materials are being substantially restructured in order to incorporate the advantages which the presence of the computer affords. The ultimate goal is to produce a new computer-based calculus text.

C4L Calculus Computers, Calculators and Collaborative Learning

Patricia R. Wilkinson	Award No: USE 8953959
CUNY Borough of Manhattan	FY 89 $ 50,000
Community College	24 months
New York, NY 10007	

The collaborative learning project is providing students, especially those from minority groups, a better chance to achieve success in calculus. The students are working in informal study groups with the assistance of specially trained tutors.

Calculus in the Liberal Arts Curriculum/ Multidisciplinary Resources for College Calculus

Ronald W. Jorgensen	Award No: USE 8953926
Nazareth College of Rochester	FY 89 $ 78,232
Rochester, NY 14610	24 months

A calculus curriculum that uses the computer algebra system MAPLE in conjunction with writing assignments that are designed to promote student learning is being developed. The courses are organized in diagnostic learning units and require students to keep a journal which is regularly evaluated by the instructor. This system of ongoing feedback between student and teacher enhances self-evaluation on the part of the student.

The Computer Revolution in Calculus: Innovative Approaches to Concepts and Applications

William E. Boyce	Award No: USE 8953904
Rensselaer Polytechnic Institute	FY 89 $ 55,000
Troy, NY 12180-3590	12 months

A new calculus course that exploits the power of a computer as an integral part of teaching and learning is being designed. Advantage is being taken of a computer's capacity to perform numerical computation, produce sophisticated graphics, and carry out extensive symbolic manipulations. Students are provided with powerful and versatile problem-solving tools and simultaneously gain a deeper understanding of the underlying mathematical concepts.

Project CALC: Calculus as a Laboratory Course

Lawrence C. Moore	Award No: USE 8953961
Duke University	FY 89 $ 198,522
Durham, NC 27706	FY 90 $ 217,773
	FY 91 $ 134,570
	FY 92 $ 39,999

Students are benefiting from a completely restructured calculus curriculum. The new curriculum features an integrated computer laboratory for exploration and development of intuition, and emphasizes writing to promote student comprehension and expression.

Calculators in the Calculus Curriculum

Thomas Dick	Award No: USE 8953938
Oregon State University	FY 89 $ 84,219
Corvallis, OR 97331-5503	FY 90 $ 75,371
	FY 91 $ 87,918
	FY 92 $ 15,000

Calculus students are benefiting from the joint effort involving universities, two- and four-year colleges, high schools, and high technology industry to develop and implement a new calculus curriculum which makes integral use of symbolic/graphical calculators. Text materials appropriate for the equivalent of three semesters of calculus are being produced and classroom-tested in a variety of instructional settings. Workshops provide continuing instructional support for teachers using the curriculum materials and symbolic/graphical calculator.

The Calculus Companion: A Computational Environment for Exploring Mathematics

Edmund A. Lamagna	Award No: USE 8953939
University of Rhode Island	FY 89 $ 161,535
Kingston, RI 02881	36 months

The calculus curriculum is being revised to provide students with more complex, real world problems, to help them develop the skills involved in performing multi-step reasoning, and to help them learn to express mathematical ideas precisely and coherently. A unique computational environment is being developed in which students use the computer as both a tutoring device and a computational aid. The system, called the Calculus Companion, consists of a user-friendly interface to the computer algebra system MAPLE and numerical computation and graphical display routines.

PROJECT ABSTRACTS: FY 1990 AWARDS

Software and Project Development for the Two-Year Calculus Sequence

David O. Lomen   Award No: USE 9053431
University of Arizona   FY 90 $ 80,000
Tucson, AZ 85721   FY 91 $ 90,000

Calculus students at many institutions are benefiting from integrated supplements: laboratory exercises, projects, problems, and software packages. The laboratory exercises are modeled after a typical chemistry or physics laboratory, projects involve the student discovering and conjecturing results, and problems are challenging and realistic. The software packages provide materials and facilities, and run on MS-DOS or Apple Macintosh machines.

Implementing Calculus Reform: Conferences, Classroom Testing, and Dissemination

Michael R. Colvin   Award No: USE 9053404
California Polytechnic State University   FY 90 $ 46,996
San Luis Obispo, CA 93407   12 months

A forum is being established for the dissemination and classroom testing of innovative ideas, pedagogy, and technological advances in teaching calculus. Classroom testing is taking place on campuses under the auspices of the California Calculus Consortium. Faculty are introduced to innovative approaches via a summer workshop, and these approaches are reinforced through follow-up activities during the academic year.

Computer Simulated Experiments in Differential Equations

David A. Horowitz   Award No: USE 9053390
Golden West College   FY 90 $ 36,000
Huntington Beach, CA 92647   12 months

Calculus students are improving their understanding of applied mathematics with the help of computer simulation programs that pictorially and graphically model real-life applications. The package includes growth and decay simulations and harmonic motion simulations.

Computer Projects and Software for the Introductory Linear Algebra Course

Gareth Williams   Award No: USE 9053365
Stetson University   FY 90 $ 24,964
Deland, FL 32720   12 months

The introductory linear algebra curriculum that is typically part of the two-year calculus sequence is being revitalized by the introduction of computer projects and related material. The projects are diverse in nature, ranging from those that explore mathematical concepts to those that involve mathematical models.

Calculus and Mathematica

J. Jerry Uhl   Award No: USE 9053372
University of Illinois   FY 90 $ 87,501
Urbana, IL 61801   FY 91 $ 90,004

Students are discovering the concepts and ideas of calculus by exploration and experimentation in a revitalized calculus course, Calculus & Mathematica, that combines the concept of calculus as a laboratory science with correct mathematical foundations. The classroom/laboratory is equipped with Macintosh computers and the text materials are presented via the Notebook feature of Mathematica.

Calculus, Concepts, and Computers

Edward L. Dubinsky   Award No: USE 9053432
Purdue University   FY 90 $ 220,000
West Lafayette, IN 47907   FY 91 $ 226,000
     FY 92 $ 200,000

Students are learning both the geometric aspects of calculus using computer graphics and the mathematical concepts via a mathematical programming language that allows them to make standard mathematical constructions using standard mathematical notation; drill and practice are being reduced by using a computer algebra system. Research into the process of learning the underlying ideas of calculus is also being conducted.

Engineering/Physical Science Second Year Calculus and Differential Equations: A Pilot Project

Leslie Hogben	Award No: USE 9053428
Iowa State University	FY 90 $ 48,241
Ames, IA 50011	12 months

Students in fourth semester calculus, differential equations, and linear algebra are benefiting from a revised curriculum that presents the underlyifig mathematics by introducing physical problems which require mathematics for their solution.

Calculus with Computing: A National Model Course

Keith D. Stroyan	Award No: USE 9053383
University of Iowa	FY 90 $ 64,000
Iowa City, IA 52242	FY 91 $ 66,000

Students in the Accelerated Calculus Program are benefiting from a new curriculum that treats beginning calculus as a laboratory course with NeXT computers and Mathematica software as the equipment. Students, in classes of 125, work on open-ended projects with assistance from graduate teaching assistants, upperclass undergraduates, and faculty.

A Reformed Calculus Program Based on Mathematics and Project CALC

William H. Barker	Award No: USE 9053397
Bowdoin College	FY 90 $ 35,000
Brunswick, ME 04011	FY 91 $ 44,000

Students are learning calculus in a discovery-based laboratory course using materials developed at Duke University and adapted for use in a liberal arts college setting. The course and laboratory materials are made available for Macintosh computers and exploit the Notebook feature of the computer algebra system, Mathematica.

Computer Algebra System Workshops, New Series

Donald B. Small	Award No: USE 9053427
Colby College	FY 90 $ 84,875
Waterville, ME 04901	FY 91 $ 10,000

College and pre-college faculty are learning how to use computer algebra systems in the teaching of calculus. Strong emphasis is placed on using these systems in such a way that the calculus curriculum is improved by introducing numerical and graphical experimentation, and by focusing on problem-solving and understanding of concepts.

A Workshop on the Undergraduate Linear Algebra Curriculum

David Lay	Award No: USE 9053422
University of Maryland	FY 90 $ 43,922
College Park, MD 20742	12 months

Mathematical Scientists are assessing the current state of linear algebra instruction in the undergraduate curriculum, laying a foundation for its improvement, and identifying priorities for further and continuing study. This endeavor is being carried out by means of a survey of linear algebra curricula, a workshop, and a conference.

Video Applications Modules in Calculus

Frank R. Giordano	Award No: USE 9053407
Consortium for Mathematics &	FY 90 $ 101,851
Its Applications, Inc.	12 months
Arlington, MA 02174	

A calculus video applications library is being produced that will expose students to exciting applications of mathematics, and includes a printed teacher's guide. The modules can also be used in faculty development activities and contain print packages with special attention to showing how to introduce these ideas into the classroom.

A Modular Calculus

William W. Farr	Award No: USE 9053430
Worcester Polytechnic Institute	FY 90 $ 56,981
Worcester, MA 01609	FY 91 $ 61,572

Students are benefiting from a new calculus curriculum that features early development of multivariable functions and derivatives, a less sequential approach to the calculus topics, and the development of team projects with computer laboratories and written laboratory reports.

Computers in Calculus, The Dearborn Project

David A. James	Award No: USE 9053385
University of Michigan	FY 90 $ 57,500
Dearborn, MI 48128	FY 91 $ 59,000

Calculus students are benefitting from a new curriculum constructed from the best curriculum development efforts around the country. After classroom testing, the results are being evaluated and a package of computer laboratory materials and an instructor's manual is being desktop-published and disseminated.

First-Year Calculus From Graphical, Numerical, and Symbolic Points of View

Arnold M. Ostebee	Award No: USE 9053363
Saint Olaf College	FY 90 $ 49,977
Northfield, MN 55057	FY 91 $ 54,965

A new curriculum combining graphical, numerical, and algebraic viewpoints on the main ideas and objects of calculus and SUPPORTED by modern computing technology is helping students understand calculus ideas more deeply and apply them more effectively.

A Model Program Using Student Research Projects in Calculus and Differential Equations

David J. Pengelley	Award No: USE 9053387
New Mexico State University	FY 90 $ 120,000
Las Cruces, NM 88003	24 months

Calculus students are benefiting from a newly designed course in vector calculus and differential equations that equips them with the basic tools of modern mathematical modeling. As a result of analyzing a sequence of discovery projects, students see how real-world questions may engender theoretical tools and how these tools may then be extended to new applications. The independent experimentation, conjecture, and testing this fosters builds the confidence needed for independent work or group leadership. Several national workshops are being held on departmental implementation of a projects-based calculus curriculum, as well as a conference on using discovery projects to teach basic ideas in calculus and differential equations.

Computer Enhancement Options for Second Year Calculus

George R. Livesay	Award No: USE 9053426
Cornell University—Endowed	FY 90 $ 93,000
Ithaca, NY 14853	FY 91 $ 97,000

Specialized software packages are being developed for the differential equations and vector and multivariable calculus topics typically included in a second-year calculus course. The new software is modeled after the MacMath and Analyzer packages already completed and classroom tested.

Developing a Projects-Based Calculus Curriculum

Stephen R. Hilbert	Award No: USE 9053416
Ithaca College	FY 90 $ 86,175
Ithaca, NY 14850	24 months

Students are achieving increased understanding of concepts, seeing the unity of the important topics in calculus, obtaining a deeper geometric understanding, and learning problem-solving skills in a new calculus course that integrates large, open-ended problems into the curriculum.

A Laboratory Approach to Calculus

L. Carl Leinbach	Award No: USE 9053401
Gettysburg College	FY 90 $ 59,225
Gettysburg, PA 17325	8 months

College faculty are learning how computer algebra systems can be used to improve calculus curricula and are designing new curricula that integrates laboratories into the new courses. The new designs being implemented are critiqued and redesigned at a follow-up workshop.

Duke University's Project CALC Test Site

Alvin J. Kay	Award No. USE 9053364
Texas A&I University	FY 90 $ 17,664
Kingsville, TX 78363	12 months

Students are learning calculus in a laboratory setting and are discovering for themselves the concepts and problem solving power of calculus. The overall framework of the laboratory and the course materials being used were developed at Duke University and are bieng tested in this setting.

PROJECT ABSTRACTS: FY 1991 AWARDS

Calculus with Computers for the Mid-Sized University: Adapting and Testing the Iowa Materials

Steven C. Leth   Award No: USE 9153277
University of Northern Colorado   FY 91 $ 45,000
Greeley, CO 80639   FY 92 $ 20,000

The Iowa materials and approach to teaching calculus are being adapted, refined, and implemented throughout the calculus sequence. A lecture approach is being integrated with an interactive computer laboratory component centered around Mathematica Notebooks. Many of the students are future mathematics techers.

Integration of Computing into Main-Track Calculus

James F. Hurley   Award No. USE 9153270
University of Connecticut   FY 91 $ 41,723
Storrs, CT 06268   FY 92 $ 79,907

The three-semester calculus sequence is being revised to integrate the computer as an active component of the learning process. A pilot program is being expanded throughout the three-semester sequence. A laboratory component is being introduced in which students will modify computer code written in True BASIC and apply the programs to a wide range of mathematical problems is being introduced.

Connecticut Calculus Consortium

Robert J. Decker   Award No: USE 9153298
University of Hartford   FY 91 $ 100,000
West Hartford, CT 06117   FY 92 $ 70,000

Students are introduced to realistic problems and to the technology (graphing calculators and microcomputers) that is capable of dealing with them. Existing materials are adapted and implemented on a state-wide basis. The laboratory materials and the text materials are integrated into the new course.

The Georgia Tech-Clemson Consortium for Undergraduate Mathematics in Science and Engineering

Alfred D. Andrew   Award No: USE 9153309
Georgia Tech Research Corporation   FY 91 $ 83,560
Georgia Institute of Technology   FY 92 $ 85,991
Atlanta, GA 30332   FY 93 $ 24,417

A large scale adaptation, refinement, and implementation project is invigorating the teaching and learning of calculus for science and engineering students. Innovations being adapted have been tested at the participating institutions. The three-year effort is taking advantage of both modern supercalculator and microcomputer technology and is incorporating group learning through team projects. The project will affect some 20,000 students over five years.

Multivariable Calculus Using Mathematica

Dennis M. Schneider   Award No: USE 9153249
Knox College   FY 91 $ 45,431
Galesburg, IL 61401   FY 92 $ 42,569

Computer technology is being exploited to produce a "leaner and livelier" multi-variable calculus curriculum supported by a collection of Mathematica Notebooks and graphics packages that provide material for classroom use as well as problems that invite students to explore further the concepts of calculus. A text is being produced which assumes that students have access to a powerful computing environment, but not necessarily Mathematica.

Disseminating Calculus in Context

James Callahan   Award No: USE 9153301
Five Colleges, Inc.   FY 91 $ 85,000
Amherst, MA 01002   FY 92 $ 120,000

The project is adapting and disseminating the Calculus in Context curriculum. The group of mathematicians with experience with this curriculum is being substantially expanded through intensive workshops, and materials are being prepared that allow instructors in high schools, two-year colleges, and four-year colleges and universities to teach the course without special training. Evaluation of the efficacy of the approach and of the dissemination efforts are part of the project.

Calculus in a Real and Complex World, Year II

Franklin A. Wattenberg   Award No: USE 9153266
University of Massachusetts   FY 91 $ 34,205
Amherst, MA 01003   FY 92 $ 29,562

A new, two-semester sequence is being developed which includes differential equations and linear algebra, and is based on the philosophy and spirit of the first-year calculus course developed under the Five Colleges Calculus in Context Project. Students gain a better grasp of the concepts of calculus when they are presented in the context of real and substantial applications that require a combination of techniques involving open-ended problems that often do not have clean, simple solutions. Writing and the use of computers are an integral part of this approach.

Calculus Reform at a Comprehensive State University with Project CALC

Charles C. Alexander   Award No: USE 9153283
University of Mississippi   FY 91 $ 60,000
University, MS 38677   12 months

The materials and approach to teaching calculus developed at Duke University are being adapted and implemented. The revised course emphasizes greater conceptual understanding through extensive writing, collaborative learning in a discovery-based microcomputer laboratory, and using mathematics for modeling real world phenomena.

Project to Adapt and Refine Purdue Model for Teaching Calculus for Liberal Arts and State Colleges

Carol L. Freeman   Award No: USE 9153259
Nebraska Wesleyan University   FY 91 $ 70,000
Lincoln, NE 68504   24 months

A consortium of institutions are adapting, refining, and implementing the approach to teaching calculus. Issues of computer anxiety and cooperative learning are being examined during the implementation phase. The adaptation phase includes introducing the use of graphing calculators.

Calculator Enhanced Instruction Project by a Consortium of NJ Community Colleges

Jean Lane   Award No: USE 9153258
Union County College   FY 91 $ 77,415
Cranford, NJ 07016   12 months

Five community colleges are adapting a calculator-based curriculum for calculus. Workshops are being conducted to introduce the faculty to the new curriculum and to begin its implementation. The faculty introduces the new approach to their colleagues as it is implemented in all calculus sections.

A Problem-Based Restructuring of Calculus

Jacob Barshay   Award No: USE 9153248
CUNY City College   FY 91 $ 55,000
New York, NY 10031   12 months

Students are working collaboratively in small teams under the guidance of advanced undergraduates on thought-provoking problems. The collection of problems is being expanded and includes materials for use with graphing calculators and guide books for the restructured course.

The Rensselaer-Albany Regional Calculus Consortium— A Curriculum Adaptation, Refinement, and Implementation Program

Timothy L. Lance   Award No: USE 9153252
SUNY at Albany   FY 91 $ 41,000
Albany, NY 12201   12 months

A revised calculus curriculum is being adapted, refined, and implemented. The faculty are attending workshops and computer-calculus classes.

Dissemination of Project CALC Methods and Materials

Lawrence C. Moore   Award No: USE 9153272
Duke University   FY 91 $ 162,165
Durham, NC 27706   FY 92 $ 97,294

Third-semester calculus materials are being completed for teaching calculus as a laboratory course. The work includes the expansion of the repertoire of classroom and laboratory projects, development of versions of alternate software and hardware environments, and completion of a high school version of the course. The methods and materials for all three semesters are being evaluated and widely disseminated by workshops for college-level and pre-college faculty, by continued publication of a newsletter, and by production of preliminary materials.

Calculus & Mathematica at Ohio State

William J. Davis   Award No: USE 9153246
Ohio State University   FY 91 $ 99,916
Columbus, OH 43210   FY 92 $ 91,277

The Calculus & Mathematica course is being extended to second year calculus. The focus is on developing of materials, testing them in the classroom, and revising them in light of the experience gained. Student outcomes are being assessed and the results reported to the community.

Complete Implementation of a Mathematica Laboratory for Calculus at a Public Metropolitan University

Richard E. Mercer   Award No: USE 9153300
Wright State University   FY 91 $ 91,629
Dayton, OH 45435   24 months

Materials are being adapted and new materials developed to implement a laboratory calculus course. The materials emphasize extensive use of Mathematica programming, especially graphics routines and conceptual questions that require written responses in paragraph form. The curricular changes emphasize a systematic treatment of the approximation of functions throughout the calculus sequence.

Implementation and Adaptation of St. Olaf First-Year Calculus in the Schools of the Chattanooga Consortium

Stephen W. Kuhn   Award No: USE 9153285
University of Tennessee   FY 91 $ 65,000
Chattanooga, TN 37403-2598   FY 92 $ 70,000

Faculty at the university, college, and high school levels are working together to improve their curriculum by adapting and implementing the model calculus program being developed. Graphical, numerical, and algebraic viewpoints are brought to bear on calculus ideas to improve students' conceptual understanding.

Gems of Exposition in Elementary Linear Algebra

Charles R. Johnson   Award No: USE 9153284
College of William and Mary   FY 91 $ 92,000
Williamsburg, VA 23185   24 months

Efforts are underway to collect and broadly disseminate gems of exposition in elementary linear algebra. These include especially insightful proofs, short and open-ended problems, longer expositional items, and machine-oriented, computational exercises, all of which are designed to communicate fundamental linear algebra ideas to beginning students. Items are being solicited from a wide range of individuals worldwide and are being published in a volume designed to be available at low cost.

Multi-HBCU Calculus Project

Walter Elias   Award No: USE 9153264
Virginia State University   FY 91 $ 50,000
Petersburg, VA 23803   FY 92 $ 100,000

The calculus curriculum is being revised. A problem solving approach that encourages experimentation and enhances the study of calculus by minority students is being implemented. The primary technical tool is the microcomputer running the computer algebra system Derive.

The Washington Center Calculus Dissemination Project

Robert S. Cole   Award No: USE 9153274
Evergreen State College   FY 91 $ 80,062
Olympia, WA 98505   FY 92 $ 145,348

Faculty from a consortium of diverse institutions are adapting, refining, and implementing approaches to teaching calculus. A select group of about twenty faculty are being trained in the new methods. This core of faculty trains faculty from twelve additional institutions, and all those involved adapt and implement one of the new approaches at their home institutions.

PROJECT ABSTRACT: FY 1992 AWARDS

Project CALC: Calculus as a Laboratory Course

Lawrence C. Moore
Duke University
Durham, NC 27708

Award No: DUE 9241916
FY 1989 $ 198,522
FY 1990 $ 217,773
FY 1991 $ 134,570
FY 1992 $ 39,999
Calculus

Students are benefiting from a completely restructured calculus curriculum at Duke University and the North Carolina School of Science and Mathematics. The new curriculum features an integrated computer laboratory for exploration and development of intuition and emphasizes writing to promote student comprehension and expression. The course materials are being developed jointly by members of the faculties of the two schools.

Calculus, Concepts, and Computers

Edward L. Dubinsky
Purdue Univ. Research Foundation
West Lafayette, IN 47907

Award No: DUE 9242137
FY 1990 $ 220,000
FY 1991 $ 226,000
FY 1992 $ 200,000
Calculus

Students are learning the geometric aspects of calculus using computer graphics and are learning the mathematical concepts via a mathematical programming language that allows them to make standard mathematical constructions using standard mathematical notation; drill and practice are being reduced by using a computer algebra system. Research into the process of learning the underlying ideas of calculus is also being conducted.

Dissemination of Project CALC Methods and Materials

Lawrence C. Moore
Duke University
Durham, NC 27708

Award No: DUE 9244194
FY 1991 $ 97,294
FY 1992 $ 97,294
Calculus

Third-semester calculus materials are being completed for teaching calculus as a laboratory course. The work includes the expansion of the repertoire of classroom and laboratory projects, development of versions of alternate software and hardware environments, and completion of a high school version of the course. The methods and materials for all three semesters are being evaluated and widely disseminated by workshops for college-level and precollege faculty, by continued publication of a newsletter, and by production of preliminary materials.

Connecticut Calculus Consortium

Robert J. Decker
University of Hartford
West Hartford, CT 0611

Award No: DUE 9244308
FY 1991 $ 100,000
FY 1992 $ 70,000
Calculus

A group of 18 institutions is working together to introduce students to realistic problems and to the technology (graphing calculators and microcomputers) that is capable of dealing with them. Existing materials are being adapted and will be implemented on a state-wide basis. The laboratory materials developed at the University of Hartford and the text materials being developed by the Core Calculus Consortium (lead by Harvard University) are being integrated into the new course.

Calculus in a Real and Complex World, Year II

Franklin Wattenberg
University of Massachusetts - Amherst
Amherst, MA 01003

Award No: DUE 9244356
FY 1991 $ 34,205
FY 1992 $ 29,562
Calculus

A new two-semester sequence is being developed that includes differential equations and linear algebra and that is based on the philosophy and spirit of the first-year calculus course developed under the Five Colleges Calculus in Context Project. Students gain a better grasp of the concepts of calculus when they are presented in the context of real and substantial applications that require a combination of techniques and that involve open-ended problems that often do not have clean, simple solutions. Writing and the use of computers are an integral part of this approach.

Calculus with Computers for the Mid-Sized University: Adapting and Testing the Iowa Materials

Steven C. Leth
University of Northern Colorado
Greeley, CO 80639

Award No: DUE 9244362
FY 1991 $ 45,000
FY 1992 $ 20,000
Calculus

The materials and the approach to teaching calculus developed at the University of Iowa are being adapted, refined, and implemented throughout the calculus sequence. A lecture approach is integrated with an interactive computer laboratory component centered around Mathematica Notebooks. Many of the students are future mathematics teachers.

The Washington Center Calculus Dissemination Project

Robert S. Cole	Award No: DUE 9244364
Evergreen State College	FY 1991 $ 80,062
Olympia, WA 98505	FY 1992 $ 145,348
	Calculus

Faculty from a consortium of diverse institutions are adapting, refining, and implementing approaches to teaching calculus that were developed at Duke University and by the Harvard Calculus Consortium. The first year involved training a select group of about 20 faculty in the new methods. This core of faculty trains faculty from 12 additional institutions in the second year, and all those involved adapt and implement one of the new approaches at their home institution.

Multivariable Calculus Using Mathematica

Dennis M. Schneider	Award No: DUE 9244434
Knox College	FY 1991 $ 45,431
Galesburg, IL 61401	FY 1992 $ 42,569
	Calculus

Computer technology is being exploited to produce a "leaner and livelier" multi-variable calculus curriculum supported by a collection of Mathematica Notebooks and graphics packages that provide material for classroom use as well as problems that invite students to explore further the concepts of calculus. A text will result that will assume that students have access to a powerful computing environment, but not necessarily Mathematica.

Calculus in Context

James Callahan	Award No: DUE 9240180
Five Colleges Incorporated	FY 1988 $ 141,707
Amherst, MA 01002	FY 1989 $ 190,845
	FY 1990 $ 174,183
	FY 1991 $ 129,392
	FY 1992 $ 74,128
	Calculus

Mathematicians from the Five College Consortium of Amherst, Hampshire, Mount Holyoke, Smith Colleges, and the University of Massachusetts are restructuring the standard three-semester calculus sequence. They are developing a new curriculum in which the four mathematical themes of optimization, estimation and approximation, differential equations, and functions of several variables are stressed. These major mathematical concepts grow out of exploring significant problems from social, life, and physical sciences. Dissemination is in the form of team-taught courses, weekend retreats, summer workshops for area faculty and high school teachers, and publication of the curriculum. These instructional materials will be used at universities, liberal arts colleges, and high schools.

A Revitalization of an Engineering/Physical Science Calculus

Elgin H. Johnston	Award No: DUE 9245079
Iowa State University	FY 1989 $ 63,600
Ames, IA 50011	FY 1990 $ 72,250
	FY 1991 $ 58,565
	FY 1992 $ 15,000
	Calculus

A four-year program is under way to revitalize the calculus course taken by science, engineering, and mathematics students. The revised curriculum stresses the modeling and problem-solving aspects of calculus and teaches students to use commercially available symbolic and numerical software to handle the technical aspects of the subject. The planning, testing, and implementation of the new curriculum is being done under the guidance of a liaison committee made up of faculty from the physical sciences, engineering, and mathematics.

Disseminating Calculus in Context

James Callahan	Award No: DUE 9245065
Five Colleges Incorporated	FY 1991 $ 85,000
Amherst, MA 01002	FY 1992 $ 120,000
	Calculus

A two-year project is under way to adapt and disseminate the Calculus in Context curriculum being taught at Hampshire, Mount Holyoke, and Smith Colleges. The group of mathematicians experienced with this curriculum is being substantially expanded through intensive workshops. Materials are being prepared that will allow instructors in high schools, two-year colleges, and four-year colleges and universities teach the course without special training. Evaluation of the efficacy of the approach and of the dissemination efforts is part of the project.

Implementation and Adaptation of St. Olaf First-Year Calculus in the Schools of the Chattanooga Consortium

Stephen W. Kuhn	Award No: DUE 9245470
University of Tennessee - Chattanooga	FY 1991 $ 65,000
Chattanooga, TN 37403-2598	FY 1992 $ 70,000
	Calculus

Faculty at the University of Tennessee, Chattanooga, the Chattanooga State Technical Community College, Southern College, and area high schools are working together to improve their curriculum by adapting and implementing the model calculus program being developed at St. Olaf College. Graphical, numerical, and algebraic viewpoints are brought to bear on calculus ideas to improve students' conceptual understanding.

Calculators in the Calculus Curriculum

Thomas P. Dick	Award No: DUE 9245080
Oregon State University	FY 1989 $ 84,219
Corvallis, OR 97331-5503	FY 1990 $ 75,371
	FY 1991 $ 87,918
	FY 1992 $ 15,000
	Calculus

Calculus students are benefiting from this joint effort involving universities, two- and four-year colleges, high schools, and high technology industry to develop and implement a new calculus curriculum which makes integral use of symbolic/graphical calculators. Text materials appropriate for the equivalent of three semesters of calculus are being produced; these materials are being class tested in a variety of instructional settings; workshops are providing instructional support for teachers using the curriculum materials and symbolic/graphical calculator.

Core Calculus Consortium: A Nationwide Project

Andrew M. Gleason	Award No: DUE 9245088
Harvard University	FY 1989 $ 346,245
Cambridge, MA 02138	FY 1990 $ 570,283
	FY 1991 $ 335,223
	FY 1992 $ 418,372
	FY 1993 $ 337,500
	Calculus

A national consortium of institutions is developing an innovative core calculus curriculum that is practical and attractive to a multitude of institutions. The consortium is led by Harvard University and also includes the University of Arizona, Colgate University, Haverford-Bryn Mawr Colleges, the University of Southern Mississippi, Stanford University, Suffolk Community College, and Chelmsford High School. The refocus of calculus is using the "Rule of Three" whereby topics are explored graphically, numeri-cally, and analytically.

Calculus and Mathematica at Ohio State

William J. Davis	Award No: DUE 9245469
Ohio State University	FY 1991 $ 99,916
Columbus, OH 43210	FY 1992 $ 91,277
	Calculus

The Calculus and Mathematica course initiated at the University of Illinois and further developed at Ohio State is being extended to second-year calculus. The focus is on development of materials, testing them in the classroom, and revising them in light of the experience gained. Student outcomes are being assessed, and results reported to the community.

Multi-HBCU (Historically Black Colleges and Universities) Calculus Project

Walter Elias	Award No: DUE 9245501
Virginia State University	FY 1991 $ 50,000
Petersburg, VA 23803	FY 1992 $ 100,000
	Calculus

Four HBCU's are working together to revise their calculus curriculum. A problem-solving approach is being implemented that encourages experimentation and enhances the study of calculus by minority students. The primary technical tool is the microcomputer running the computer algebra system Derive.

The Georgia Tech-Clemson Consortium for Undergraduate Mathematics in Science and Engineering

Alfred D. Andrew	Award No: DUE 9245528
Georgia Institute of Technology	FY 1991 $ 83,560
Atlanta, GA 30332	FY 1992 $ 85,991
	FY 1993 $ 24,417
	Calculus

A large-scale adaptation, refinement, and implementation project is invigorating teaching and learning calculus for science and engineering students. Innovations being adapted have been tested at each of the two participating institutions and at Iowa State University, New Mexico State University, and Cornell University. The three-year effort is taking advantage of both modern supercalculator and microcomputer technology and incorporates group learning through team projects. The project will affect some 20,000 students over five years.

Integration of Computing into Main-Track Calculus

James F. Hurley	Award No: DUE 9245548
University of Connecticut	FY 1991 $ 41,723
Storrs, CT 06268	FY 1992 $ 79,771
	Calculus

The three-semester calculus sequence is being revised to integrate the computer as an active component of the learning process. A pilot program begun in 1989 is being expanded throughout the three-semester sequence. A laboratory component is being introduced in which students will modify computer code written in True BASIC and apply the programs to a wide range of mathematical problems.

The Rensselaer-Albany Regional Calculus Consortium—A Curriculum Adaptation, Refinement, and Implementation

Timothy Lance   Award No: DUE 9246707
SUNY - Albany   FY 1991 $ 41,000
Albany, NY 12201   FY 1992 $ 16,333
                                                    Calculus

The University at Albany, in collaboration with Rensselaer Polytechnic Institute, is incorporating interactive adaptation and refinement of two computer intensive approaches to teaching calculus; the Albany and Rensselaer models from previous years; ongoing training workshops about this learning environment for our own mathematics faculties and those of area two- and four-year colleges and secondary schools; creation of a "virtual classroom" for broader dissemination of our ideas and testing of a model for distance learning of mathematics in a computer-intensive environment; site testing of existing NSF-sponsored computer calculus initiatives. A goal is to create a distributed version of our own local computer classrooms.

Calculus and Mathematica

Jerry Uhl   Award No: DUE 9252484
University of Illinois - Urbana   FY 1992 $ 150,000
Urbana, IL 61801   FY 1993 $ 150,000
                                           FY 1994 $ 150,000
                                                    Calculus

Calculus and Mathematica is a laboratory course in calculus based on electronic interactive notebooks written within the Mathematica system. In two years, the teaching of Calculus and Mathematica has spread to more than 20 colleges and 6 high schools. The proposed work is continuing this project by extending the development, dissemination, and evaluation of the existing Calculus and Mathematica project, together with a pilot development of a Differential Equations course.

Fully Renewed Calculus at Three Large Universities

Keith D. Stroyan   Award No: DUE 9252486
University of Iowa   FY 1992 $ 80,471
Iowa City, IA 52242   FY 1993 $ 39,751
                                           FY 1994 $ 42,184
                                                    Calculus

This collaborative project is being implemented at the University of Iowa, University of Wisconsin - La Crosse, and Brigham Young University. Renewed calculus materials developed at the University of Iowa are being be revised, tested, and refined at these three institutions. In addition to learning traditional calculus skills, students are exposed to new ideas through large open ended projects on a variety of scientific and mathematical projects.

The Rhode Island Calculus Consortium Module Project

Lewis Pakula   Award No: DUE 9252468
University of Rhode Island   FY 1992 $ 150,966
Kingston, RI 02881   Calculus

The Rhode Island Calculus Consortium is composed of Rhode Island university, college, and high school faculty who are seeking to introduce new approaches to calculus instruction. It is adapting ideas and materials from successful pilot calculus projects to create a series of self-contained instructional modules for use in the first two semesters of college or high school (AP) calculus. Each module is organized around a set of problems, projects, and examination items, as well as text material and class and group activities. Modules will be piloted, revised, and evaluated by consortium instructors. A fundamental feature of this project is that the modules will be assembled by the consortium members themselves. A Summer Institute and a Calculus Colloquium will disseminate to a wider regional audience the products of the project effort, and with them. the broad aims and ideas of calculus reform.

The University of Connecticut Computer-integrated Calculus Project

James F. Hurley   Award No: DUE 9252463
University of Connecticut   FY 1992 $ 177,746
Storrs, CT 06268   FY 1993 $ 197,064
                                           FY 1994 $ 219,434
                                                    Calculus

Connecticut's program seeks to (1) involve the computer as a tool for fostering in-depth conceptual understanding of both the ideas and techniques of calculus; (2) promote analytical thinking in dealing with quantitative relationships; (3) sharpen students' geometric, numerical, and theoretical intuition, and increase their ability to visualize associated basic relationships; (4) convey the usefulness of and provide practice with the computer as a tool for carrying out mathematical computations; (5) develop in students the habit of working cooperatively with peers to analyze and solve problems, and the confidence to persist in mathematical analysis of complex, multi-step phenomena; and (6) make calculus reform feasible at institutions where cost and ease of transition from a traditional calculus program are significant issues. Dissemination to 78 participating high schools in the University's Cooperative Education Program is under way

Metrolina Calculus Consortium: Implementing a
Technology-based Calculus Curriculum

Marv K. Prichard   Award No: DUE 9252502
University of North Carolina - Charlotte FY 1992 $ 153,484
Charlotte, NC 28223   Calculus

The Metrolina Calculus Consortium is designed to facilitate
the implementation of different calculus reform materials
into schools in the Charlotte, North Carolina, area thorough
workshops and related activities. The project includes four
components: adaptation of curriculum materials to fit local
needs, faculty development, research on student learning,
and dissemination. The project provides support and
resources for teachers at these institutions to adapt and
implement a technology-based calculus curriculum.
Research on student learning is a critical component of this
project.

Implementation of Calculus Reform at a Comprehensive
State University with Project CALC

C.C. Alexander   Award No: DUE 9252516
University of Mississippi   FY 1992 $ 145,941
University, MS 38677   Calculus

The University of Mississippi is implementing a program of
calculus reform using materials and methods of Project
CALC. By the conclusion of the proposed phase-in period,
all regular faculty in the department will have an opportunity
to teach a section of Project CALC. The university will
serve as a calculus test site for the Mississippi Alliance for
Minority Participation as the alliance strives to transform
"gatekeeping" courses into "gateway" courses. One project
goal will be to refine the Project CALC materials, and
possibly the implementation strategy, based on our
experiences (this will include the adaptation of the materials
to MathCAD 3.1).

Preparing for a New Calculus

Anthony L. Peressini   Award No: DUE 9252475
University of Illinois - Urbana   FY 1992 $ 91,394
Urbana, IL 6180   Calculus

The conference/workshop "Preparing for a New Calculus" is
bringing together 80 leaders in mathematics curriculum
reform, including calculus reform, school mathematics
reform in light of the NCTM Curriculum Standards, and
educational technology initiatives. The deliberations of this
conference will yield a set of action oriented
recommendations focusing on the implications of the
emerging calculus courses on school mathematics training,
and on how recent developments in content and methods in
high school mathematics will impact the calculus courses.

Calculus Reform in Western Appalachia: A Consortium
Approach

James H. Wells   Award No: DUE 9252494
University of Kentucky   FY 1992 $ 378,459
Lexington, KY 405060057   FY 1993 $ 324,081
                                     FY 1994 $ 47,460
                                                Calculus

Mathematicians at five public and four private institutions in
three western Appalachian states (Kentucky, Tennessee, and
Virginia) are developing and integrating modem calculus
curricula into their instructional programs. In addition to
systematically involving departmental colleagues in the
teaching of the new curricula, they are developing and
implementing training programs for teaching assistants and
workshops to familiarize faculty in other disciplines with the
philosophy and implications of the revisions. The organized
system of collaboration and information sharing among the
participants is integrated into an extensive program of
dissemination, including television, directed at colleagues in
sister institutions and high schools. They are placing
particular emphasis on efforts to inform and involve high
school teachers of calculus and precalculus.

Mid-Atlantic Regional Calculus Consortium

Joshua A. Leslie   Award No: DUE 9252508
Howard University   FY 1992 $ 75,000
Washington, DC 20059   Calculus

MARCC—a consortium of five Black universities, a two-
year community college, and two inner-city Black high
schools—is proposing to implement the Harvard Core
Calculus Course. MARCC enrolls over 2,000 minority
students in its Calculus I and II courses.

Calculator-Enhanced Instruction Project by a Consortium
of New Jersey and Pennsylvania Educational Institutions

J. Lane   Award No: DUE 9252491
Union County College   FY 1992 $ 229,734
Cranford, NJ 07016   Calculus

This two-year project addresses reform of the mathematics
curriculum, particularly in calculus and precalculus, by
incorporating the use of graphics and symbolic manipulation
calculators to enhance teaching and learning following the
Clemson model. The consortium will include seven two-year
colleges, one private four-year college and 3 high schools.
Faculty will attend intensive, hands-on weekend workshops
on the TI-81 and HP-48 calculators in early fall 1992.
Participants will convene again for four additional weekend
workshops scheduled throughout 1993 and 1994.

A Reformed Calculus Program Based on Mathematica and Project CALC

William H. Barker
Bowdoin College
Brunswick, ME 04011

Award No: DUE 9249589
FY 1990 $ 35,000
FY 1991 $ 44,000
FY 1992 $ 3,878
Calculus

Students are learning calculus in a discovery-based laboratory course using materials developed at Duke University and adapted for use in a liberal arts college setting. The course and laboratory materials are made available for Macintosh computers and exploit the Notebook feature of the computer algebra system Mathematica.

Calculus Reform Workshops

Donald B. Small
Mathematical Association of America
Washington, DC 20001-0000

Award No: DUE 9253119
FY 1992 $ 66,313
Calculus

Three Calculus Reform Workshops were held during the summer of 1992. Each five-day workshop had 26 participants and included the following: an overview of calculus reform projects, extensive participant work with specific reformed approach, experienced curriculum reformers as workshop instructors, participant development of curriculum materials, and establishment of support network among participants.

Implementation and Dissemination of the Harvard Consortium Materials in Arizona, Oklahoma, and Utah

David Lovelock
University of Arizona
Tucson, AZ 85721

Award No: DUE 9252521
FY 1992 $ 181,365
FY 1993 $ 297,031
FY 1994 $ 321,604
Calculus

Our goal is to have substantial implementation of the Harvard Calculus Consortium materials throughout Arizona, Oklahoma, Utah, and the surrounding regions by the end of the project period. This will be done in two complementary steps. First, over the full three years, the coalition (Arizona State University, Brigham Young University, Northern Arizona University, Oklahoma State University, and the University of Arizona) will implement reform calculus. Second, during the last two years, we will expand our coalition to include satellites, which are other two- and four-year institutions and high schools from the region eager to implement reform calculus. The effort will include preliminary discussions, series of workshops, in-site visits, and electronic networking.

A Video on Using Supercalculators in Curriculum Reform

John W. Kenelly
Clemson University
Clemson, SC 29634

Award No: DUE 9252524
FY 1992 $ 60,000
Calculus

A video on using supercalculators as one approach to curriculum reform has been designed, developed, tested, and evaluated. The video illustrates the changes that take place in mathematics classrooms when personal supercalculators are used regularly for instruction, homework, and tests. The completed video was mailed, free-of-charge, to each of the nation's 2,5000 mathematics departments in fall 1992.

Workshops for Dissemination of Calculus Reform Projects

A. Wayne Roberts
Macalaster College
Saint Paul, MN 55105

Award No: DUE 9252529
FY 1992 $ 163,515
FY 1993 $ 168,520
Calculus

We plan to conduct 16 one-week 24-person workshops about calculus reform projects, 8 in the summer of 1993, 8 more in the summer of 1994, at sites across the country. Workshops will be run by leaders in the calculus reform movement who will describe their projects and how they overcame obstacles (need for new resources, skepticism of client disciplines, colleague resistance) that confronts any curricular reform. Participants will be asked to consider their own situation and to formulate a plan for action in their home institution.

A New Calculus Program at the University of Michigan

Morton Brown
University of Michigan - Ann Arbor
Ann Arbor, MI 481091220

Award No: FY 1993 $ 324,081
FY 1992 $ 175,000
FY 1993 $ 125,000
FY 1994 $ 100,000
Calculus

The University of Michigan is redesigning its first-year calculus curriculum in content, delivery, and style. The principal features of the new program are cooperative learning, greater use of technology, increase in geometric visualization, new syllabus emphasizing problem solving, and quantitative reasoning. Starting from some pilots sections, all calculus sections will be reformed by fall 1994. There will be an extensive training program for course instructors, including a handbook and supplementary materials for use along with the Harvard Consortium text.

Calculus and the Bridge To Calculus

Mulloy Robertson	Award No: DUE 9252501
Volunteer State Community College	FY 1992 $ 49,984
Gallatin, TN 37066	Calculus

Volunteer State Community College is collaborating with area high schools and vocational schools in analyzing all math courses and their outcomes as a basis for recommending curricular changes that could result in higher-level math skills for High School graduates. Two workshops are planned The first will identify the problems, and analyze and discuss them. A goal is a 25% reduction of students who need developmental math from these specific high schools.

Fully Renewed Calculus at Three Large Universities

John Unbehaun	Award No: DUE 9253958
University of Wisconsin - La Crosse	FY 1992 $ 59,009
La Crosse, WI 54601	FY 1993 $ 30,178
	FY 1994 $ 26,113
	Calculus

The University of Wisconsin - La Crosse is revising, testing, and refining the renewed calculus materials developed at the University of Iowa. The University of Wisconsin - La Crosse will fully implement in all calculus classes these calculus materials. We believe that this reformed calculus approach will better prepare our students to pursue careers in science and technology by being proficient at basic calculus skills, while also learning computing with Mathematica and applying the ideas of calculus in a variety of large and small open-ended projects.

Fully Renewed Calculus at Three Large Universities

Gurcharan S. Gill	Award No: DUE 9253959
Brigham Young University	FY 1992 $ 18,235
Provo, UT 84602	FY 1993 $ 32,225
	FY 1994 $ 33,540
	Calculus

This collaborative project is being implemented at the University of Iowa, University of Wisconsin - La Crosse, and Brigham Young University. Renewed calculus materials developed through the NSF supported project at the University of Iowa is being revised, tested, and refined at these three institutions. The project involves extensive training of faculty and teaching assistants to use these materials and approaches. In addition, ideas are introduced through large open ended projects on a variety of scientific and mathematical problems.

PROJECT ABSTRACTS: FY 1993 AWARDS

The Western Pennsylvania Calculus Technology Consortium

Frank Beatrous
University of Pittsburgh
Pittsburgh, PA 15260

Award No: DUE 9352874
FY1993 S 144,443
Mathematics

A technology enhanced calculus courses is being established at two campuses of the University of Pittsburgh. The approach adopted In the University courses is based on the Calculus & Mathematica (C&M) project developed at the University of Illinois and at Ohio State University. The CALC-TECH project will facilitate large scale implementation of the C&M project on the two University campuses. This will require workshops for faculty and graduate teaching assistants, development of materials, and evaluation

Maricopa Mathematics Consortium Project

Alfredo G. de los Santos
Maricopa County CC District
Tempe, AZ 85281

Award No: DUE 9352897
FY1993 $ 100,000
Mathematics

The Maricopa Mathematics Consortium (M C)--composed of the Maricopa County Community College District, Arizona State University, and four public school districts in Maricopa County--are instituting a two-year project that is resulting in significant systemic change in the teaching/ learning process in precalculus mathematics. The participating institutions, who serve 235,000 students, have a long history of cooperation and joint development. This collaborative effort includes restructuring the curriculum, developing materials to reflect the changes to be made, and using technology and new pedagogies. Faculty/staff development is being provided during all phases to build faculty support and confidence in these new approaches to teaching mathematics (776 faculty). During the evaluation process, they are examining outcomes at key benchmark points in the areas of faculty development, student learning, and developmental processes.

Development Site for Complex, Technology-Based Problems in Calculus

Brian J. Winkel
Rose-Hulman Institute of Technology
Terre Haute, IN 47803

Award No: DUE 9352849
FY1993 $ 100,000
Mathematics

Faculty and students at Rose-Hulman Institute of Technology in cooperation with several high school teachers are developing and disseminating complex problems in calculus. These technology-based problems are interdisciplinary in nature and integrate concepts from calculus, engineering, and physics. The project will produce Mathematica resources to accompany problems, and writing up guidelines for using problems in high school and college calculus settings. Suggestions for nonMathematica users are being offered in each situation as well as guides on how to use the materials.

Mathematica Laboratory Projects Projects for Calculus and Applied Mathematics

William Barker
Bowdoin College
Brunswick, ME 04011

Award No: DUE 9352868
FY1993 $ 77,871
Mathematics

Building on Bowdoin College's three years of experience as the Mathematica laboratory development site for Project CALC, this project is to design and construct Mathematica computer laboratory notebooks for multivariable calculus and applied mathematics. The primary focus of the development efforts is on applications from disciplines which are new to the calculus and applied mathematics curriculum. Dissemination will involve a workshop as well as electronic networks. These notebooks are distinguished by their complete development of topics under study, their focus on models from other disciplines, and their use of graphics routines to develop geometric understanding.

Interactive Modules For Courses Following Calculus

Lawrence C. Moore
Duke University
Durham, NC 27708

Award No: DUE 9352889
FY1993 $ 181,827
FY1994 $ 180,860
Mathematics

The authors are developing interactive laboratory and classroom materials to support follow-on courses for students completing reformed calculus course: linear algebra, differential equations, and applied mathematical analysis. A seven-member development team, working at five different institutions over a two-year period, will develop approximately 70 interactive text modules. Each module will be developed in one of the systems Mathcad 3.1, Maple V, or Mathematica, and translated by student assistants to each of the other two systems. This project will build on and reinforce the acquired concepts and shared experiences of students moving on from reformed calculus courses. These students know how to work in a lab, how to work with partners, and how to work on their own. Moreover, they have had introductions to many of the important ideas that will be developed further in the follow-on courses. Such students will find conventional courses less than satisfactory as preparation for their scientific or engineering careers in the 21st century.

Full-Scale Calculus Revitalization and Evaluation Project

Joe A. Marlin   Award No: DUE 9352845
North Carolina State University   FY1993 $ 99,878
Raleigh, NC 276958208   Mathematics

An experience-based, application-driven, two-year calculus sequence involving 50 faculty members, 50 graduate teaching assistants, and over 4,000 undergraduate mathematics, science and engineering majors each semester is being developed. The three year plan includes to revitalizing undergraduate calculus curricula and instructional techniques through curriculum enhancement, experiential computer laboratories, and intensive faculty development. The project demonstrates innovative adaptations of small-scale models previously developed by others, in addition to a new set of curriculum enhancements emphasizing applications in other disciplines, which may be adopted or further adapted at institutions of any size.

Implementation of the Harvard Core Calculus at Stony Brook

Anthony V. Phillips   Award No: DUE 9352843
SUNY-Stony Brook   FY1993 $ 155,864
Stony Brook, NY 11794   Mathematics

The Harvard Consortium calculus is being implemented throughout the calculus courses, from beginning precalculus/calculus through multivariate calculus, in the context of a thorough revision of our first-year and remedial classes. This involves adapting the Harvard curriculum to meet the needs of a diverse student population, who fall into three categories: the 1600 students a year who take the first term of the Precalculus/Calculus sequence, for whom the precalculus end of the curriculum needs to be expanded; the 900 who take either the slow-stream or the mainstream calculus, who can use the curriculum essentially in its present form, and a smaller number of well-prepared students, who took some calculus in high school, and would benefit from enrichment material to supplement the more routine sections of the curriculum.

A Three Semester Integrated Calculus/Physics Sequence

Wesley Ostertag   Award No: DUE 9352841
SUNY Dutchess County College   FY1993 S 95,010
Poughkeepsie, NY 12601   Mathematics

The investigators are writing laboratory materials for and team teaching an integrated three semester sequence of courses in calculus and physics. The core of the project is the creation of a series of mini-labs which develop analytical topics in tandem with physical principles using data-gathering equipment connected to personal computers. The investigators are considering the long-standing pair of problems in introductory science education: applications meant to motivate the calculus are often developed poorly and/or out of context by the calculus instructor, and mathematical tools needed in the physics course are often used by the physics instructor before they have been adequately developed in the calculus course.

Mid-Atlantic Regional Calculus Consortium

Joshua Leslie   Award No: DUE-9352865
Howard University   FY1993 $100,000
Washington, DC 20059   FY1994 $100,000
                                       FY1995 $100,000
                                       Mathematics

This project is being funded as part of the Alliances for Minority Participation project, a consortium effort initially involving Howard University, Hampton University, Morgan state University, and the University of the District of Columbia. The goal of HAMP is to double within five years the number of individuals from minority groups underrepresented in science, engineering, and mathematics who obtain bachelor of science degrees in these fields. The calculus activities, centering around the Harvard Consortium reform text, will complement the strong student support which will be provided through HAMP in achieving this goal. Plans are also outlined for developing a precalculus course consistent with the Harvard Consortium materials. Student working groups modeled after the PDP approach.

Calculus, Linear Algebra, and ODE's in a Real and Complex World

Franklin Wattenberg   Award No: DUE 9352828
University of Massachusetts-Amherst   FY1993 S 79,871
Amherst, MA 01003   Mathematics

Being continued is the development, dissemination, and assessment of an integrated two-year sequence to replace five courses--Calculus I, Il, and III, Linear Algebra, and Ordinary Differential Equations. The motivating idea driving the project is that these subjects should be taught in the context of real and engaging problems. By studying these subjects in context, students are better able to understand them and to use them outside the mathematics classroom. Emphasized are very substantial and realistic problems that require the application of a combination of mathematical techniques and that are open-ended, often without clean, simple solutions. Computers and writing are essential parts of the approach. There is a synergy between writing and meaningful problems that engages students on a high intellectual plane requiring them to understand and articulate the mathematics, the applications, and the connections between them.

Large-scale Calculus Revision at Penn

Dennis M. DeTurck　　　　Award No: DUE 9352824
University of Pennsylvania　　　　FY1993 $100,000
Philadelphia, PA 19104　　　　Mathematics

A full-scale revision of the entire Calculus program is being undertaken making extensive use of student projects drawn from scientific and business disciplines. The three main thrusts of the revision are (1) substantial reconstruction of the first and second year Calculus syllabi; (2) emphasis upon collaborative learning techniques; and (3) use of computation as a vehicle to encourage collaborative learning. The emphasis is on designing and implementing means of presenting new approaches to Calculus effectively on a large scale. Materials are being adapted from existing Calculus reform efforts. In addition, materials and support are being developed for faculty and students getting used to using computers to do mathematics and to incorporating problems from other disciplines.

The Washington Center Calculus Dissemination Project

Robert S. Cole　　　　Award No: DUE 9352900
Evergreen State College　　　　FY1993 $252,506
Olympia, WA 98505　　　　Mathematics

The Washington Center for Improving the Quality of Undergraduate Education, is augmenting and extending for another two years the work of its Washington Center Calculus Project. The goals for the project are to build sustainable institutional commitment to calculus reform efforts already initiated, to deepen understanding of assessment tools appropriate to some of the new pedagogies being used in teaching calculus, to increase the number of institutions using reform calculus curricula, and to document the dissemination model they are using for statewide calculus reform. This proposal broadens the scope of the work in which they are currently engaged, and focuses more upon sustaining and institutionalizing curricular reform, rather than initiating it, and upon developing and documenting assessment and dissemination methods appropriate to regional initiatives.

Differential Equations: A Dynamical Systems Approach

Paul R. Blanchard　　　　Award No: DUE 9352833
Boston University　　　　FY1993 $117,260
Boston, MA 02215　　　　FY1994 $102,241
　　　　Mathematics

A large-scale revision of the traditional sophomore-level ordinary differential equations course is being developed which emphasizes qualitative theory with a distinct dynamical systems orientation. A discussion of difference equations will precede the introduction of differential equations, the computer will be used to analyze solutions, and a more detailed discussion of nonlinear systems will be included. A consortium of colleges and universities will assist in developing the syllabus and materials, as well as serve as initial test sites for the textbook.

Calculus For Comprehensive Universities and Two-Year Colleges

Gregory D. Foley　　　　Award No: DUE 9352894
Sam Houston State University　　　　FY1993 $ 70,000
Huntsville, TX 77341　　　　Mathematics

The goals of this project are to design and implement a curriculum for the first year of calculus at a set of eight comprehensive universities and two-year colleges. The program adapts and synthesizes methods from successful calculus reform efforts and is crafted to meet local needs. The curriculum stresses and facilitates cooperative learning, develops visual thinking with the aid of interactive graphing technology, and uses writing to help students learn and communicate mathematics. These methods are integrated in an environment that focuses on the central ideas of calculus and provides a progression of problems and projects to challenge students while improving their confidence and study skills. A volume of student assignments will be compiled to serve as a resource for college faculty across the nation.

Bridge Calculus Consortium Based at Harvard

Deborah Hughes-Hallett　Award No:DUE 9352905
Harvard University　　　　FY1994 $141,847
Cambridge, MA 02138　　　　FY1995 $259,810
　　　　FY1996 $299,212
　　　　FY1997 $299,135

The Calculus Consortium based at Harvard University, with funding from the National Science Foundation, has developed, tested, and disseminated an innovative single variable calculus course. In this project, the effort is being expanded to include precalculus and the second year of calculus. A major component of the proposed project is dissemination. The dissemination effort for the proposed project is being modeled on the workshops, minicourses, newsletters, test sites, and networking that have proved successful in the current project.

PROJECT ABSTRACTS: FY 1994 AWARDS

Interactive Electronic Third Semester Calculus Laboratory Materials for Personal Computers

Thomas F. Banchoff   DUE 9450721
Brown University   FY1994 $181,650
Providence, RI 02912   Mathematics

Third semester calculus of two and more variables can often be taught most effectively when classroom lectures and texts are supplemented by interactive computer graphics laboratories. The laboratory materials already developed at Brown University for introductory differential geometry using a network of SUN workstations are now being redesigned, using recent electronic book technology, so that it can run on Macintosh level computers. This is making the visualization software available to a much larger range of institutions and students, in engineering, natural sciences, life sciences, and economics, as well as computer science and mathematics.

A Computer Based Introductory Differential Equations Course

Robert L. Borrelli   DUE 9450742
Harvey Mudd College   FY1994 $ 250,127
Claremont, CA 91711   Mathematics

The aim of this project is to design a syllabus, and modules to support that syllabus, which adds computing and mathematical modeling as essential components for the introductory ordinary differential equations (ODE) course. Many modules are be self contained units whose main focus is a topic involving the construction and analysis of a mathematical model using a computer based approach. Some modules are designed to shed new light on traditional topics found in introductory ODE courses. The material is being produced and field tested by a Consortium of seven institutions and is being made available to the authors of the next generation of textbooks on differential equations. The overall administration of this project is at Harvey Mudd College, but the work will be done at all seven Consortium institutions: Cornell University, Harvey Mudd College (CA), Rensselaer Polytechnic Institute, St. Olaf College (MN), Washington State University, Stetson University, and West Valley College (CA). Building on the network and experience accumulated during a previous DUE grant on faculty enhancement, the Consortium is broadening the scope of its newsletter CoODEoE to publish information about the new ODE course syllabus and modules before, during and after test-runs. In addition, CoODEoE is disseminating information about other ODE projects around the country. A textbook publisher is adding additional support for workshops, computing programming expertise, and evaluation.

Reforming Calculus Instruction in Puerto Rico

Rafael Martinez Planell   DUE 9450758
U of Puerto Rico Mayaguez   FY1994 $197,935
Mayaguez, PR 00709   Mathematics

Through this project, a coalition of two and four year colleges and universities in Puerto Rico are changing the way calculus is taught. In addition to preparing faculty to use the Core Calculus Curriculum materials, the project is promoting and assessing the use of collaborative learning techniques with the Hispanic student population. The materials produced as well as the assessment of student achievement will also be of value to universities with large Hispanic populations.

A Comprehensive Calculus Project for Comprehensive Universities

Curtis C. McKnight   DUE 9450760
U of Oklahoma   FY1994 $ 209,992
Norman, OK 73019   Mathematics

Reformed calculus materials developed by Ostebee and Zorn at St. Olaf College are being used in the teaching of first year calculus at The University of Oklahoma. Faculty in the Department of Mathematics are extending this model of calculus reform by undertaking tasks representing adaption, assessment, faculty and graduate student development, components for pre-service teachers, and dissemination. The St. Olaf first year calculus model is being adapted and extended to fit a large, comprehensive university setting including the range of difficulties implied by this (large lecture with recitation, GTA taught classes, honors classes in an open admission university, etc.).

The project includes the development of a model for GTA training along with supporting materials and training videotapes, a model for faculty development, including materials, suitable for sustaining such a calculus program, and careful assessment of student performance. Reports on assessment and evaluation will be useful to other comprehensive universities seeking similar reforms. An important component addresses pre-service secondary mathematics students: adaptation of the GTA training materials for prospective mathematics teachers; training experiments involving pre-service students; participation in GTA training sessions; and, observations both in university classes and in a local high school which has adopted the St. Olaf's model for calculus. Dissemination plans include packaging materials for dissemination, production of written materials, and visits and locally held workshops. At least one monograph will be prepared for publication.

A Regional Center for Calculus Reform at Northeastern University

Terence J. Gaffney   DUE 9450764
Northeastern University   FY1994 $ 299,973
Boston, MA 02115   Mathematics

A regional center for calculus reform is being established in the Boston area centered at Northeastern University. As part of this effort, curriculum reforms are being made to the multivariable and differential equation courses, and throughout the full range of calculus and precalculus courses. The completed reforms will affect over 150 sections; over 15,000 students will enroll in these courses in the next 5 years. Working with Northeastern's Comprehensive Regional Center for Minorities, the project is strengthening and helping create precalculus and calculus programs in public high schools in the City of Boston, and other urban school districts with large minority enrollments. As an incentive for students and teachers to join this program, Northeastern will award credit to all high school students completing a calculus course through this program. There will also be a regional network of school teachers and college faculty committed to the adaptation and dissemination of the ideas and techniques of the reform movement, reaching thousands of area students. This network will maintain a software and materials library and sponsor seminars, workshops, and regional conferences.

Multivariable Calculus from Graphical, Numerical, and Symbolic Points of View

Arnold M. Ostebee   DUE 9450765
Saint Olaf College   FY1994 $ 150,061
Northfield, MN 55057   Mathematics

A curriculum and materials development project focusing on multivariate calculus builds on and extends the work by the PI's on single variable calculus. Faculty at St. Olaf College are writing, field testing, and disseminating complete course materials, including a textbook, for a rethought multivariable calculus course that emphasizes the development of conceptual understanding and analytical reasoning skills. The course and its supporting materials share a pervasive mathematical theme and a consistent pedagogical strategy: to combine, compare, and move among graphical, numerical, and algebraic viewpoints on calculus. Graphical, numerical, and symbolic computing technology is used to support this emphasis. The materials use technology freely but are not bound to any specific hardware or software platform. The technology requirements are defined generically in terms of functionalities, rather than by brand name. The text assumes that students have access to technology that has at least the capabilities of a CAS. Specific applications to a particular CAS can be found in supplements and lab manuals.

West Point Core Curriculum Conference in Mathematics

Donald Small   DUE 9450767
US Military Academy   FY1994 $ 95,000
West Point, NY 10996   Mathematics

The CUPM Recommendations for a General Mathematical Sciences Program detailed a new view of undergraduate mathematics, stating a philosophy centering on goals for student learning and outlining a needed broadening of the curriculum. Subsequent CUPM documents call for a 4 semester sequence of study that contains the core of mathematics normally found in three semesters of calculus and semesters of discrete mathematics, linear algebra, differential equations, and probability and statistics. This call for a broader view of mathematics dovetails well with the vision for secondary school mathematics outlined in the NCTM's Curriculum and Evaluation Standards. A workshop during summer, 1994, brings together leaders from secondary, community college, liberal arts and research university mathematics to discuss such 7-into-4 programs and to review their implications for the variety of individuals and client disciplines they serve.

The United States Military Academy at West Point has a fully implemented model of a 7 into 4 curriculum. The activities of the workshop focus on discussing the content that should be central to each of the seven curricular areas, examining the Academy's program, and considering how content can be better integrated to better prepare the students served. The products of the conference consist of proceedings with commissioned papers and reactions, details on the West Point program, and actions and recommendations of the workshop. Panel discussions based on the workshop will be proposed for future MAA meetings.

Gateways to Advanced Mathematical Thinking: Linear Algebra and Precalculus

Al Cuoco   DUE 9450731
Education Development Ctr   FY1994 $ 236,029
Newton, MA 02160   FY1995 $ 273,177
                                        FY1996 $ 272,977

This project is building on previous research and curriculum development work to develop flexible understandings of topics in precalculus, calculus, and linear algebra. Exemplary course materials will be produced using a modular approach to package the concepts and activities. Based on the broad mathematical themes, these materials will make use of constructivist pedagogies, involving cooperative learning, computer technology, and alternatives to traditional lecturing. Field testing of the curriculum will provide sites for research into the way students learn the topics and environments for teacher enhancement.

Graphing Calculator Resource Materials for Calculus and Pre-Calculus

W. Frank Ward   DUE 9450744
Indian River Cmty College   FY1994 $ 100,000
Fort Pierce, FL 33454   Mathematics

Faculty in the Department of Mathematics at Indian River Community College are developing instructional materials for the use of the graphing calculator to support teaching with the Harvard Calculus Consortium materials that are appropriate for a two year college audience. Materials are being developed for both teachers and students. The teacher supplement contains examples and instructions on how and when to effectively use a graphing calculator as an instructional tool. Less emphasis is placed on traditional lecture and more emphasis is placed on exploration and group problem solving. The material integrates practical applications taken from astronomy and physics and other areas that use mathematics. Students generate real data through the use of a Calculator Base Laboratory System (CBL). The device enables the collection of real time data for analysis with the graphing calculator. The collection and analysis of real data provides the means to integrate practical laboratory experience into pre-calculus and calculus. The instructional materials include the use of the CBL system.

The student supplement is designed with an emphasis on using technology to explore mathematical concepts and write about the findings. Projects address four types of activities: (1) "get started" activities to familiarize students with using the calculators; (2) activities to help students understand concepts; (3) projects to promote deeper understanding of concepts that may be used as group projects or for advanced students; and, (4) investigation projects for topics beyond what is discussed in class. This range of activities will benefit the slower student by providing activities for understanding as well as the advanced student by providing enrichment material. The conceptual approach combined with real applications, emphasis on understanding and problem solving benefits all students.

Instituting Calculus Reform: A Community College State University Consortium Model

William L. Lepowsky   DUE 9450735
Peralta Cmty Col Dist Off   FY1994 $119,703
Oakland, CA 94606   Mathematics

The Peralta Community College District, San Francisco City College, California State University Hayward, and San Francisco State University are forming a consortium to jointly revitalize instruction in calculus by adapting the Harvard Calculus Curriculum. Faculty from each institution are jointly developing the curriculum which will incorporate instructional methods and strategies that have proven to be effective in increasing the success rate of calculus students. The results are being published in a

Calculus Reform Handbook. The respective mathematics departments are working to institutionalize the developed reform curriculum in all calculus classes at the four institutions. In addition, the core group are developing a preparation program for mathematics faculty who have not participated in the curriculum development effort.

Calculus, Concepts, Computers and Cooperative Learning: Assessment and Evaluation in Terms of Dissemination Goals

Edward L. Dubinsky   DUE 9450750
Purdue Univ Research Fdn   FY1994 $50,000
West Lafayette, IN 47907   FY1995 $168,000
                                                        Mathematics

This project is engaging in assessment, evaluation and research into how students learn mathematical concepts to determine effects of the calculus reform curriculum and dissemination activities from the previously supported Calculus project at Purdue. Assessment activities will be closely related to research into how students learn mathematical concepts and will also include studies of the effect of the role of the teacher when changed from instructor to facilitator and the use of group learning. Current dissemination approach will include the preparation of an Instructor's Resource Manual, workshops and the continued growth and development of a network of implementers. A model will also be developed for evaluation of student attitudes, performance, and conceptual understanding in comparison with students who took calculus in standard models.

Assessing the Workshop Calculus with Review Project

Nancy H. Baxter   DUE 9450746
Dickinson College   FY1994 $ 100,505
Carlisle, PA 17013   Mathematics

Workshop Calculus with Review I and II is a unique sequence of courses which integrates a review of fundamental concepts needed to explore ideas in calculus with the study of concepts normally encountered in Calculus I. Materials have been developed with support from the NSF ILI program and the U.S. Department of Education FIPSE program. The goals of this two-year assessment project are to: (1) identify a list of "community norms" - that is, a list of key concepts students who take an integrated pre-calculus/calculus sequence are expected to know; (2) design and implement internal and external assessment tools to evaluate student learning gains and attitudes, with a particular emphasis on the impact of technology; and, (3) track student retention rates, continuation rates, and performance in subsequent mathematics classes and classes in other disciplines that have a calculus prerequisite. Materials are being beta tested in a variety of settings and developers are assisting colleagues in adapting the materials for use at their institutions. Based on outcomes of the assessments undertaken by this project, the Workshop Calculus with I and II instructional materials will be refined and published for widespread dissemination. The results of the project will be published and the assessment tools will be distributed.

Workshop Calculus with Review I and II is a integration of pre-calculus concepts with Calculus I material and is designed for students who are not prepared to enter Calculus I. The sequence provides a gateway into the traditional calculus sequence. In the Workshop environment, lectures are replaced by interactive teaching, where students learn by doing and by reflecting on what they have done. In the workshop environment, lectures are replaced by interactive teaching in which no formal distinction is made between classroom and laboratory work. In class, students explore mathematical ideas on their own and discover their own approaches to solving problems, using technology as appropriate and by working collaboratively on tasks in their Student Activity Guides.

Sag